The Autobiography

of

A · PIONEER GIRL

LOU CURTIS FOSTER

Designed and Produced
by

B. C. LANGLEY
Chicago, Illinois

MARY LIZZIE
1853 - 1944

Prelude

Not all were fair
But of the blots
I shall not write.
'Tis better they're forgot
And shut away from human sight.

They were but trifles,
Bits of temper,
Quick retort,
A moment's disobedience
Soon atoned, forgiven, and forgot.

It may be - who shall dare deny?
Almighty Love will wipe away the stain,
And leave all fair and white again.
I pray it may be so.

                              L.C.F.

Preface

I have written these stories of early life to please my children. They found them more fascinating than 'Mother Goose'. The first little daughter, while still a toddler, would climb in my lap and demand, "Tell me about when you were a little girl, Mama," and so on down through the four. It was always, "Tell about when you were a little girl." So many times I ransacked my memory for the odd bits of life, the little adventures, even the every day doings that my toddlers found so interesting, that the telling became a sort of schooling to fasten many things in my memory otherwise forgotten.

I told them the stories my mother and her sisters, Vina and Libbie, had told me of life on the Hudson River, when they were children; stories of the trail when the Bennett family pioneered into Western New York, of Lyman, of the wedding, and of that eight day journey by stage-coach from Ontario County, New York, to Morelands, Kentucky. I knew personally and loved Grandfather Jacob Bennett and Grandmother Caroline Valentine Bennett. Through the story telling I grew to know almost as well, Great Grandfather Jacob Bennett and his wife Barbara Brower Bennett. To the child who listened to these stories Jacob Bennett and his wife Barbara Brower were real people. They still stand before her as distinct personalities, undimmed by the years of three quarters of a century.

Father Lyman Curtis was a silent man, strict, even stern to his children. Grown, they reverenced him, but his little children stood in too great awe to climb freely upon his knee and demand a story. Lyman's two younger sisters, Elinor and Mary, and the two boys for whom he stood guardian, Henry Clay and Wilbur Fisk, were the story tellers on the Curtis side. They were all good story tellers, for in the old days when books and magazines were few and newspapers still fewer, the art of talking was well established, and story telling filled in the wide spaces of written history. We grew up with vivid pictures of western Massachusetts, of the Hudson, and that first Catskill Manor House that had been built of tile brought over from Holland by an ancestor. Perhaps we knew more and remembered better than had we gained our knowledge from books.

As they grew into womanhood, these daughters of mine kept up their insistent demand, but now it was changed to, "Write them down, Mama, so that they may not be forgotten, that your grandchildren may know as we have known, the bits of life from the times of their Great, Great Grandfather and Grandmother Curtis."

This I have done to the best of my ability. If other children find them interesting, I shall be gratified. The straight line illustrations are the work of a daughter, who has for years so illustrated her letters to the children, to their great delight.

CONTENTS

MARY LIZZIE
PRELUDE
PREFACE

APPENDIX ... THE WAR SONGS

CHAPTER ONE

Mary Lizzie's Background

The Beginning

In the beginning God made Heaven and Earth
Then he sent forth the life bringing wind to give birth
Forests rose, flowers bloomed.
And the wind rushed on to people the earth.

L.C.F.

ON her father's side Mary Lizzie traced back to royal English blood. Elizabeth St. John, sister to Oliver St. John, Chief Justice of England in Cromwell's time, married one of her ancestors and, early in 1600, came with him to New England. This man and one other of her ancestors, graduates one of Cambridge and one of Oxford, were nonconformist ministers who came to New England to establish a purer form of religion. They were among the founders of Harvard College, and one or both served on the first Board of Directors.

On her mother's side her line traced back to William Prince of Orange. Anneke Jans was his granddaughter. According to tradition she married Roeloff Roeloffson, who was not of royal blood. For this, she and her heirs to the seventh generation were dis-inherited by her royal grandfather. Her inheritance, left in trust to be administered to later generations, was kept intact. It was a royal sum for which these descendants from the seventh generation down were fighting. Anneke came to New Amsterdam with Roeloff Roeloffson in 1633. Here he obtained a grant of land, but did not live to enjoy it. On his death she married again.

Through this second husband, Evardus Bogardus,

1

A PIONEER GIRL

Mary Lizzie's ancestors entered the line of descent and became heirs of Anneke Jans. Evardus Bogardus was the first minister to New Amsterdam and came to America in 1633. With his marriage to Anneke he assumed, aside from his clerical duties, the cares of a landed proprietor. Her grant of land practically covered all of Long Island and the adjacent coast. Wall Street was built upon this land. Trinity Church stands on it. According to record, the land west of Broadway extending to the Hudson River, and from the Battery past Park Row, was leased to the trustees of Trinity Church for a period of ninety-nine years. At the expiration of this term it was returned to the heirs, and re-leased by them to the Church, but at the expiration of the second term it was not returned, and a suit to obtain possession was instituted by the heirs.

The records show that the property, originally belonging to Holland, passed from the State into the hands of William Prince of Orange "as a reparation and satisfaction." Records are few, but the tax receipts bear out these traditions and show that the descendants of Anneke Jans Bogardus actually occupied these lands until 1785, when representatives of Trinity Church forced them out, burned the fences and destroyed the crops, leaving desolation everywhere.

There is an old map in the Hall of Records in New York City that shows an agreed boundary line for the Jans Bogardus lands. There are also records to show in effect that these lands were under lease to the Trinity Church Corporation both before and after the Queen Ann Grant of 1705.

The Will of Anneke Jans Bogardus, recorded in the Clerk's office at Albany, New York, was written in Dutch and dated January 29, 1663. Her farm lay on Long Island. It was about sixty-two acres, and the boundaries were given by the Bogardus heirs in their bill of complaint against Trinity Church Corporation. Such were the

2

records produced at the trial of Bogardus versus Trinity Church. But Wall Street was powerful, and added to its power was the sanctity of the Church. The heirs were not willing to give up their rightful inheritance, so the fight for ownership went on and on and on, the ancestors of Mary Lizzie among them.

Jacob Bennett the First was married in Trinity Church to Barbara Brower Bogardus, who was descended from the Prince of Orange and was an heir to Anneke Jans of Wall Street and Trinity Church fame. This first Jacob Bennett was a private in the war of the Revolution and was taken prisoner and held six months on a prison ship at Quebec. At the close of the war he returned to his farm near New York City, where he spent the remainder of his life. He and his wife Barbara were buried in the little cemetery for which Great Grandfather Bennett gave the ground. He never claimed the bonus given by the State of New York to soldiers.

Grandfather Jacob the Second, their son, was a splendid specimen of pioneer manhood. He was six feet tall and well proportioned, with curly black hair and dancing eyes, and was a leader in the choir and singing-school. He was a handsome young giant in the days when he went courting dainty Caroline Valentine. He had a rich baritone voice, and until very old was the choir leader whereever he lived.

That old time singing-school was social as well as musical, a merry gathering of the people of the neighborhood. The leader, tuning fork in hand, took his stand in front. A sharp rap of that little instrument on the table or against his teeth brought forth a note that his voice took up. Do-re-me-fa-sol - up the scale he ran, beating time until the right note was reached, when every voice in the room took it up and the singing-school was opened. Happy faces and merry voices made the old time singing-school what the song proclaimed it to be, -"beau-ti-fule".

3

A PIONEER GIRL

The song follows:

> Oh the Singing-School's beau-ti-fule
> Oh the Singing-School's beau-ti-fule
> If I had you for my teacher
> I should be a happy creature
> For I dote upon the Singing-School.
>
> Childhood joy is very great
> A-swinging on the garden gate
> But others think the pleasure more
> A-sliding down the cellar door.
>
> Oh some think this and some think that
> But all agree there's greater satisfaction
>   to be always had
> At Singing-School, as I have said.

Caroline Valentine was a slip of a girl weighing less than one hundred pounds, pink and white, with blue eyes and brown hair, when she married Jacob and was carried off to his cabin in the big woods on the eastern side of the Hudson.

When Caroline was ten years old, she had gone to live with a family on the west side of the cascades at Platter-kill, to be a companion for their little daughter, and the two grew up as sisters. It was from this home that she was married, and her adopted parents outfitted her royally, as a daughter should be, with feather bed and pillows, home-spun blankets, hand-woven counterpanes and bed and table linen. Here is a word picture that has come down to us of Caroline in her wedding gown, and very pretty she must have looked. It was of white swiss with short tight Dutch waist, low neck and little puff sleeves. The skirt was full, rather short, and perfectly plain in front, all the fullness being gathered in the back. She

4

wore under it skirts - many of them - so tight she could hardly sit down, but they were necessary to hold out the full skirt of the dress.

Their wedding took place on the twenty-fourth of February, eighteen hundred and nineteen. It was late enough for the ice in the Hudson to be broken up, and the river was full of floating cakes. Grandfather Bennett, accompanied by his best man, made the crossing in a skiff in the water, wherever possible; between waters, dragging the skiff out onto the ice and carrying it to the next open stretch; while Caroline and her friends on the opposite bank anxiously watched their every movement. It was a hazardous trip, and one to which Jacob was not willing to submit his bride; hence the longer way home, around by crossings that had greater stability.

The road, hardly more than a trail in those years, led down the west side of the river to the village of New York. Here the Hudson was crossed on a barge; then followed another trail up the eastern side to the new home. Thirty miles of river and forest, with an ever present fear of lurking Indians, was what the little bride faced when the goodbyes were said and Jacob swung her up to the padded seat behind him. Swung her up with little slips and catches, his booming laugh a response to her, "Now, Jacob, don't you drop me." At New York they crossed over to Brooklyn, and rested and visited among the relatives and friends. Then on the trail once more, up the river to the lonely cabin in the woods of the east side. They were young and happy, and it was all a part of the big adventure.

The poor little bride was so afraid in that lonely cabin in the woods that, as night came on, she would sit on a flat stone near the door and wait until Grandfather Jacob closed his shop and came singing down the path to catch her up and carry her into the house. Once, in those pitiful forest years, she found her 'live coals', left uncover-

5

ed, had become black and dead. There was but one re-
course. She must snatch up her iron coal kettle and has-
ten off through the woods to a neighbor, the nearest one
being more than a mile distant, and beg for a new start
of 'live coals'. A fire she must have to get supper. Poor
little Caroline ran, lest night and darkness overtake her
under the shadow of those fearful trees. The coals were
secured and kept covered carefully, for they were more
precious than silver or gold, until her trembling fingers
put them into a nest of kindling. Then the bellows fanned
slowly, and at last the flames were dancing. Never again
did she forget to cover her coals.

Caroline runs with her coal kettle for a
fresh start of 'live coals'

The loneliness of the big woods grew upon her, and so
homesick did she become that she once walked the thirty
miles to her father's house, thereby losing the first little
life that would have brightened those somber woods.

Grandfather Jacob was seven years apprenticed to a
blacksmith, but when the time of his apprenticeship was
over he was a master at the trade. His wage at the close

of these seven years consisted of a new suit of clothes, tools for a blacksmith shop, and a small sum of money. He held for years a government contract to shoe the army mules at West Point. This was a big order, for he found it necessary to take two helpers with him. The government contract was renewed again and again.

West Point was our first military school. Washington recommended it, and as far back as seventeen hundred and seventy-six Congress ordered it established, but it was not until eighteen hundred and two that the Military Academy of the United States at West Point began its existence with ten cadets under Major Jonathan Williams as its first contingent. There were practically no examinations either physical or mental for these cadets, and they were permitted to enter at any age or at any time in the year. In this modest way began our present complete system of military education.

Jacob had his shop by the side of the river road, and there were many among the comers and goers who needed his help. Steadily life grew about the Bennetts. The trail became a well-marked road. Another trail opened out to the West, and a few hardy spirits passed over it -- hunting, trapping or seeking adventure. 'Westward Ho' seemed to have been the slogan ever since the world started moving. Jacob caught the fever, and after a few years the cabin in the woods was left dark and silent while the Bennetts followed that alluring trail and joined the 'Westward Ho' crowd.

While still a young man, Grandfather Bennett had charge of the building of a little church that stood on a bluff over-looking the Hudson River. It was erected for a mission chapel, and so used for many years. In eighteen hundred and seventy-four, long years after the building had been done, Grandfather Bennett, returning to his home in Maysville, Missouri from Santa Fe, New Mexico, found hanging on the old adobe walls of a ranch house at which he stopped for the night, a painting of his little mission church, so perfect that he knew it instantly. The owner verified Grandfather's knowledge, and when he was ready to leave in the morning he was given the picture. Unfortunately this picture was destroyed in the fire that burned Uncle Corty's house.

Grandfather Jacob Bennett saw Fulton's first steamboat, 'The Clermont', as it made its maiden trip up the Hudson River. Crowds gathered and all cheered its passing, even though many of them had been foremost among the doubting ones.

Grandfather was a great joker, and always ready with a witty reply and a jovial laugh. Both he and Grandmother Bennett were fond of backgammon, and it solaced many an hour as they grew older. At these times Grandfather was full of his pranks, and would pick up a 'man' and put it down where he wanted it, but never once did it pass unnoticed. Grandmother's quick, decisive, "Now, Jacob, you put that back," would bring a boom of hearty laughter, and the 'man' was put back in its place. They were such genial companions that the whole family, in leisure moments, would always be found around them.

Caroline and Jacob raised eight children, the eldest of whom was Jennett, who was to be Mary Lizzie's mother. Following her came Nehemiah, Cornelius, Joseph and William, all of whom were born at Cold Springs on the Hudson before the move west. Elizabeth (Libbie), Lewis, Vina (Amanda Malvina), and Courtland or Corty, were born in Shelby Township. Grandmother Caroline's children called her by the endearing name of 'Little Mother', and spoke often of her beautiful hands. Such capable hands they must have been, for she bore nine children, and with the help of her dear eldest daughter, Jennett, reared eight of them to healthful manhood and womanhood.

Nettie, as the little girl was called, was present at the opening of the Erie Canal in eighteen hundred and twenty-five. She was only four years old at the time, and her father lifted her to his shoulder that she might see over the heads of the crowd. She never forgot that meeting of the waters, the splash, the toss of the white foam, and the wild cheering of the people.

She heard the 'Town Crier', whose duty it was to call the people to church and to cry 'lost and found articles', and liked to watch him as he passed up and down the streets of little old New York, calling the people to church with the steady, insistent beat of his drum and his compelling cry, "Oyez, Oyez, all ye good people, come to the meeting house." Once she was part of the wild excitement when a little girl was lost, and she heard the beat of the drum as he passed up one street and down another, with his far reaching cry, "Oyez, Oyez, all ye good people, help to find the lost child." Shops were deserted and work was suspended, as the people turned out to search. Then later she heard the glad quick roll of the drum that told the child was found.

Trinity Church and Wall Street where Jacob Bennett the first married Barbara Brower Bogardus

CHAPTER TWO

Going   West

IN eighteen hundred and thirty-three when Jennett was twelve years old, the Bennetts moved from Cold Spring on the Hudson into Western New York, Orleans County and settled in Shelby Township, near Medina. In those days that was moving west - far west. Of the eight children who followed Jennett, Willie, the fifth child was a baby when the move was made, and the only record of the trip concerned him: "Willie's cap blew into the river." No mention was made of the danger of crossing the Hudson River on flat-boat or in skiff, nor of the long road that wound up and over the Palisades; no mention of the great Tonawanda Swamp nor of the Tonawanda Indians who dwelt therein. Only this funny little skit: "Willie's cap blew into the river." No other record was needed to form a picture. Caroline's quick, "Stop, Jacob, Willie's cap blew off," halted the oar in midair. It scooped up the floating cap and tossed it into the boat before joining the rower's rhythmic sweep again.

Over the river, the whole outfit - cut down to barest necessities - was transferred to a lumber wagon drawn by horses; oxen were too slow. The children were tucked in wherever a cranny was found large enough to hold one. Caroline, with baby Willie in her arms, was made

10

as comfortable as possible, and the long road out to the west began.

Of the ache that wrung Caroline's heart when she turned for a last look at her beloved river, no one ever knew. She had need for courage, and God gave it to her in unstinted measure.

There followed days of travel over a road yet scarcely more than a trail, and when nights came and the children were packed into the closest possible bed space at some little road house, Caroline was only too glad to lie down anywhere so she might rest.

The Post Office was at Millville; Lockport was the metropolis of the region. It was to Lockport that the Bennetts went on shopping expeditions, when an article was required that could not be produced at home. Linen underwear from the flax was made at home. Wool from the sheep's back was carded and spun into dresses for the women and coats and trousers for the men. Stockings were knitted. Linens for table and beds, comforters and blankets were all of home manufacture. Even ordinary cobbling was done by Jacob. Only Jennett and her mother had the luxury of 'boughten' shoes. It was a struggle for existence; yet they were healthy and contented.

Jennett danced off to school over the winter snow banks in a dress and hooded cloak of crimson merino, the rich coloring well suited to the fresh pink of her face, the sparkling gray eyes, and the mass of blue-black hair. That crimson merino cloak was wrapped around her slim young form when she threw herself down in the fresh snow and stretched out her arms full length, to make her 'picture'; then came the quick upward whirl, this way and that, to shake herself free from the snow. With hood untied and cloak open, she was a picture quaint as well as pretty.

Below the full gathered skirt were the pantalets. Those funny pantalets were not panties, just pantalets, a cover-

11

ing for the knee, worn for modesty or ornament in summer, for warmth in winter. They came some inches below the dress, straight across, and were embroidered or lace edged. Jennett's winter ones were like the crimson dress and were gathered into a band, ruffle edged, that buttoned around her ankles.

The little girl had been growing up fast. She was her mother's assistant and bore the care of the children almost equally with her. She combed hair, washed hands and faces, put the little flock into clean clothes, and marshalled them off to the meeting house in Medina. Church going was a habit with these pioneers, and was never omitted. The Bennetts were Methodists, and Methodists in those days were straight laced and severe as to the outward forms of religion.

Once when sixteen year old Jennett donned a new spring bonnett of plainest straw, tied demurely under her chin, she wore it happily to church, conscious of the single rose on the outside and the bit of bright ribbon on the inside that framed her fresh young face. Her happiness was rudely shattered by the minister, who eyed her with stern disapproval and commanded: "Jennett, go home and take off that flower. Take out that ribbon. They are the lure of the devil." Then he preached a blood-curdling sermon on this subject. A tearful Jennett went home and, counseled by her mother, obeyed the stern command. But her obedience was doubtless accompanied by many an inward 'swear', ill becoming a Methodist maiden of that day.

About the same time that the Bennetts settled in western New York, the Curtises, from the Berkshire Hills of Massachusetts, also settled in Shelby Township. The two farms were near each other, and Lyman and Jennett, then in their teens, went to school together. Lyman was the oldest in a family of fourteen children, six girls and eight boys.

12

Newman Curtis, Lyman's father, had but one wife, Maria Van Bergen, but he jokingly called these fourteen children his two families, because, beginning with Catherine and Lyman, they ran a girl and a boy, a girl and a boy, down to the middle of the line; then there were two boys, Newman and John, who broke the order, to be taken up again by Martha and Edwin and followed to the end, where Henry and Wilbur finished it off.

Catherine and Lyman were great playfellows. Once their mother interrupted a most spectacular play. Catherine was all but buried, apparently with her full consent. "I was planting her for seed to grow another little sister," was sturdy little Lyman's prompt reply to his mother's horrified questioning.

A log cabin sheltered the big family until just before the birth of Wilbur, last of the fourteen children, when they moved into a new home called the Curtis Manor House, that for many years was the finest in the vicinity. It was built of cobble stones from the size of a goose egg down to a bird's egg, all brought by wagon from Lake Ontario, eighteen miles away. The small stones were used for decorations, forming circles all over the house. The doors and window sills were of cut stone.

Lyman and Jennett attended Milville Academy, which must have ranked high, for both were well educated not only as to essentials but in the more advanced subjects as well. After they finished, both became teachers. When Lyman was about twenty-two he went to Kentucky and founded a private school. Jennett taught in New York until Lyman came back after her.

They were married in eighteen hundred and forty-three, on the Bennett farm in front of the fireplace that filled one end of the big living room. Very winsome the bride must have been, with her smooth raven black hair, fine gray eyes and fresh pink skin that never knew any other powder than that which came through the sides of a flannel starch bag. Her dress was a dove-colored silk, full skirted and shirred into the tight whale-boned waist, with wide, flowing sleeves. It was finished with a deep collar of embroidery at the neck. Underneath she wore several starched underskirts, each finished with a three inch layer of solid embroidery, the bride's own work.

Another delightful picture of Jennett is given us in one other of her bride dresses, a challis, very soft and fine, with purple flowers on a white ground. Like the wedding dress, it was full skirted. The perfectly plain waist was whale-boned in bodice shape and laced down the back so close and tight that one brother would hold it squeezed around that trim waist while the other brother drew it up.

The groom was no whit less attractive. He was tall and straight, with piercing blue eyes, aquiline nose, and dark hair. His beardless face, with its square set chin, showed strength of character. The straight upstanding hair that gave him so much distinction was the result of painstaking effort when, as a small boy, he nightly wet it and brushed it and banded it back until it was trained. It made a marvelous framing for the highbred, austere

14

face that held such strong character lines. All through his life he wore the standing collar and black silk stock or cravat that was the fashion at that time.

CHAPTER THREE

Kentucky

The eight day trip to Kentucky

AFTER the wedding Lyman and Jennett went by stage direct to Morelands, Kentucky. It took them eight days to make the trip. Happy years followed for the young couple. Together they taught in the private school Lyman had established. One summer there was an epidemic of cholera and many deaths followed. Both Lyman and Jennett were down with the dread disease, but through good nursing and watchful care they were soon able to get around again. The following spring school had to be closed while measles went the rounds.

Early in their stay Lyman bought near Ashland, Henry Clay's plantation, an attractive little home set among roses and acanthus bushes, for which he paid eleven hundred dollars. Lexington was the center of the finest region in Kentucky, and Lyman and Jennett had highbred Kentuckians for neighbors and friends. Nehemiah, Jennett's brother, visited them and taught a term of school while there. He was a great favorite in the community.

16

After he left, Lyman carried on an uninterrupted corre-
spondence with him for many years. They were fun loving young people, and in one of his letters Lyman described a merry Christmas party:

And well we may be happy, for we have had an
unusually brisk Christmas; yes, and don't you think
Jennett hasn't made a party, which came off Christmas Day. We had a goodly number present, among
whom were most of your friends in this section.
And I must tell you one thing Jennett did on the occasion of which I am sure you will not approve.
What do you think it was? Why, she made some
egg-nog. I told her she should not do it, but you
know she will sometimes do as she pleases, and so
she did on this occasion. But what I doubt not you
will think was much worse is that, after it was
made, the old man himself had to take a little!

Jennett's egg-nog was a delicious winey drink much
served in those old Kentucky days, of the sort that cheers
but not inebriates. It brightened stale hours and helped
life to flow easily.

This same brother Nehemiah had taken a course in
Spencerian penmanship in New York City while still a
very young man. Could you see the framed specimen of
his writing that hangs in Mary Lizzie's home, you would
not wonder at the open admiration of the Kentuckians, -
"Your last letter to me made the people stare, you may
be sure. The P.M. says to himself when he took it out
of the bag, 'That's from Bennett, I'll bet,' so everybody
about must needs take a peep at it. Now why do you think
they all thought it was from you? I'll tell you the reason
they gave, when I asked them how they knew. 'Because,'
said they, 'there is nobody in the U.S. besides can write
like it.' I thought it quite a compliment to you, and one

17

A PIONEER GIRL

,meant in good earnest. "

The happy years here were shadowed only by the death of little 'Newmy', the first son. He lived only two years, but long enough for the grandparents on both sides to enjoy the baby sweetness of this first grandchild, for Lyman and Jennett went back with him on a visit to New York that lasted while Lyman taught school in a room of the Curtis house. Schools were few and far between and teachers were scarce, so Lyman's help was needed here for the younger children. Soon after the death of 'Newmy', Caro was born, and she grew strong and fine. Once when Henry Clay swung her up into his arms, she hid her nut-stained hands behind her and bravely piped up, "I'm my Daddy's bruty, I am."; a fact that remained a fact through her life.

Lyman and Jennett never owned slaves, but it was imperative that this young mother and teacher should have help. Since only 'niggers' were to be had, they were hired from one of the patrons of the school, paying for their hire directly to the owner.

Jennett went to Lexington with Lyman to make purchases. She bought dresses for herself and for Caro, and a sash to match, and when they were bought the little girl felt very much dressed up as she went walking with her Daddy.

These Northerners were adopted and loved by the warm hearted Kentuckians, who were loved in return. In proof whereof they named their second baby girl after two Kentucky belles, Mary Harp and Lizzie Gossip. When she was asked her name, she always gave it as "Mary Lizzie Harp Gossip Curtis."

Mary Lizzie was not beautiful then or ever after, just a pink and white mite of humanity tipping only six pounds when dressed. Her head, with its satin-smooth black hair, was small enough to slip into a teacup. In the day-

18

time it was covered with the cunningest of lace caps, all beruffled and trimmed with pink ribbon bows. For the rest they put next her body only linen: linen bands, linen diapers, linen shirts - sleeveless save for the turn-over band of fine drawn work that edged the whole top. Her tiny feet wore the softest of white wool bootees, and were folded inside a pinning blanket of fine white flannel finished with a band of feather stitching. Next came the flannel skirt of fine white wool, long enough for the beautiful silk embroidery that scalloped the bottom to show when Baby Girl could be lifted into someone's arms. Over this was the under-skirt of fine linen, its edge simply finished with hand-made lace. Over all went the sheerest linen-lawn, sleeves tied at the wrists with baby ribbon, the full skirt finished at the bottom with a wide hem and bands of tucks, each tiny one laid perfectly with the thread of the goods, put in by patient fingers with the finest of needles. Of course the skirts were full, shirred in carefully laid stitches, and they were long, l-o-n-g. The bath and dressing were a daily ceremony, and when it was finished they folded around her a square of silk and wool embroidery.

Yet she thrived and grew, even as the baby of today, in its simpler dressing, grows. The little limbs grew straight and true and the back waxed strong. Oh, she was a little aristocrat, worthy of her royal ancestry; she might even have inherited the Kentucky dower of beauty, had her parents been Kentuckians, but unfortunately for Mary Lizzie's looks they were not.

She was born when Henry Clay was at the height of his career, and with her first breath she drew in pure Republican doctrine, for he, the first great Republican, was a neighbor and a friend of her father during their life at Morelands. Often did she hear the old campaign refrain:

A PIONEER GIRL

> The man that I give my hand to
> Must be the firm friend of 'Old Clay'.

One of Lyman's treasures was a cane given him by Henry Clay himself. It was mounted with ivory head, silver name plate, and brass ferule complete, and was presented to Father Lyman when he was leaving Kentucky for the west. This cane was adopted by the small brother, Georgie, years later, as his favorite 'horsey', and ridden by him and became worn and broken off until it was only about two-thirds its original length; whatever became of this priceless treasure Mary Lizzie has no knowledge.

Some interesting Extracts from Letters written by Lyman to friends between 1844 and 1846, after Clay's defeat in the Presidential race.

From the first, in February of 1845: "But really, one would suppose you had gotten into some serious difficulty from the tone of the first part of your letter, but before I read it through I was happy to find it was only the explosion of some extra gas which had accumulated during the furious heat of the late Presidential canvass, or perhaps you, as a matter of precaution, found it necessary to let off some steam as our Whig ship ran aground, to prevent a more fatal explosion. Our noble Whig ship aground, did I say? No! Never while she is manned by the Gallant Tars of New England, need we fear such a disaster. But rather say, she for a while is disabled by the whirlwinds of slander and detractions which she has lately encountered. But she will soon be under way, prepared to battle for our constitution and our Country's honor.

The Whigs as a party stand proud and glorious in defeat, and who would not rather suffer defeat while holding up the hands of our patriot, Henry Clay, than triumph with the filth of a Tory spawn. There are many causes which have combined to produce our defeat, among the most prominent of which are illegal voting and double dealing among our opponents... Mr. Polk was held up to be the particular and especial friend of a special interest in our portion of the country, and in another declared to be its uncompromising opponent. I cannot dwell upon this trait of that party without trembling for the liberties of our country. If the people can be thus deluded now, while the graves of our fathers are fresh before us and while the famous address of a Washington is still ringing in our ears, and while the parting word of a beloved and lamented Harrison is still calling upon them to 'Remember the true principles of the government'; I say that, if they can be thus deluded now, when all this is before them; what are we to expect in the future?"

In another letter to another friend, they discuss Mr. Clay's speech of February 7, 1839, wherein his opponents charged him with presenting amalgamation as the remedy for slavery. Lyman deduces the meaning of this speech of Clay's as follows: "As I understand, it is simply this: that considering the much greater relative increase of the white race over the black, they, the whites, would in the course of time become so numerous as to take the place of the blacks, whose labor would be more expensive compared with that of the whites in proportion as their number was relatively lesser, and as the labor of intelligent men is and must always be preferred to that of the thick skulled African, the slave holder as a matter of pecuniary interest would dismiss his slaves one after the other until slavery was unknown amongst us, and thus Mr. Clay would show by logical reasoning that slavery is destined to become extinct by the inevitable laws of population."

Lyman quotes from the same speech, wherein Mr.

21

Clay says, "I am, Mr. President, no friend of slavery...
Whenever it is safe and practicable, I desire to see every
portion of the human family in possession of it (civic
liberty), but I prefer the liberty of my own race to that of
any other race."

Lyman continues: "Now, sir, after reading such sen-
timents as those for his own race, how was it possible
that you would charge Mr. Clay with a desire to stain the
fair face of his own species with negro blood?...Gener-
ations to come may ask how Henry Clay was defeated, and
the answer will be borne back in these words, 'Fraud,
lying and misrepresentation'."

Lyman was a loyal Clay man - but he knew and accepted his defeat. He does not in any of these letters of 1844-
1846 express any fear that the Whig party may also become defunct.

In another letter to another friend he discusses the
Tariff question, and incidentally he questions that friend's
definition of Republicanism and proceeds to "Settle some
preliminaries as to what Republicanism might or might
not be defined to mean". Particularly is the new party
opposed to the Tariff, or as Lyman puts it, "Goes in for
a nullification of the Tariff". Wool is the article under
discussion, and Lyman discusses it pro and con from the
Whig standpoint. He finishes with, "Your foundations
were laid in sand and will not stand the test of facts and
reason. The truth is you have mistaken the whole intention of the Whig policy as to protection. The Whigs do
not go for protection solely for the small amount of loss
or gain it produces on a definite article absolutely considered. Ah, no, sir, our object is to make America happy and free."

In one letter to a friend who has asked his opinion
upon the founding of a new party: "You wish to know my
opinion in regard to Native Americanism, causing a new
organization of the political parties of the day...upon

22

that subject I will just say that. I am opposed to any new
organization of the Whig party. That party is essentially
American in all its measures ... 'My Country may she
ever be right, but right or wrong, my Country.' I care
not what party is in power, if war overtakes us it is our
duty as patriots to lay aside party animosities and all
unite to bring it to a successful and speedy termination."

I regret to say that we have no copies of these letters
later than June of 1846. But in this last letter he repudi-
ates the charge of "abolitionism ... as ungenerous as it
is untrue".

When the new Republican party was organized, Lyman
was in its ranks.

It was not permitted Henry Clay to live to take a part
in the great struggle of the '60s between the North and
South. He died in July of 1852. In one of Lyman's letters
to a brother-in-law, he writes of attending the funeral.
The letter follows:

Funeral services commenced at 9 o'clock A. M.
The city is full of strangers, and still they are
coming from every quarter. All the public buildings
in the city are draped in mourning. Military companies from various parts of the state march with
arms reversed. Everything wears a sad and sorrowful aspect. All parties, sects, and orders of
men unite and seem to vie with each other in ren-
dering homage to the memory of the dead. We are
situated in the upper story of a large business
house, being kindly invited to take possession for
the time being by the proprietor; hence we have a
commanding prospect of all that passes in the street
below, and as the funeral procession will pass the
entire length of this street, we would not possibly
have a better situation to see. As I write, all the

bells of the city are tolling forth their solemn chimes, and the cannon is speaking in thunder tones of the mighty and illustrious dead. The procession, or rather a part of it, consisting of the Committee of Senators who brought the corpse from Washington and others from different parts of the Union are now proceeding with slow and mournful tread to Ashland, to take charge of the <u>Dead</u> and return it to the city, where the procession <u>will</u> be joined by the Military, citizens and strangers generally who care to take part in it.

The remains of Mr. Clay arrived in this city from Louisville last night at seven o'clock and were immediately conveyed to Ashland to be mourned over there by the aged partner of his joys and sorrows. While at Ashland, and before the Committee take possession of the body, there will be a private funeral service performed according to the Episcopal Order, a member of which denomination Mr. Clay was.

While the procession has gone to Ashland, let us pause for a few moments and consider who has fallen, thus to cause the whole nation to put on mourning. 'Henry Clay', than whom none has, since the days of 'the Father of his Country', done more for his Country. His name is written upon every page of our Country's history in letters of gold for the last fifty years, and not only is his name familiar to the people of the great North American Republic but throughout the world. Whereever man has struggled for that liberty which we possess, and which his voice was ever raised in eloquent tones to maintain, there will the voice of mourning be heard, and ascend with our own to that God who in His mercy has brought this great affliction upon us as a nation. I pray God that He will

24

raise up others who may take his place on the watch tower of Liberty and raise their voices at the approach of danger.

But who now living can take the place of Mr. Clay in our national counsels? There are none who have the influence that he had, and none who can direct with such unerring wisdom. But I must close these reflections, as those stationed at the window announce that the procession is approaching with muffled drum and measured tread, silent and sad.

First comes the Chief Marshall, followed by a cannon taken in the Mexican War, manned by the gallant fellows who accomplished the proud act, all in deep mourning--cannon shrouded and flag furled. Second, several companies of Cavalry, four abreast, horses covered with black. Third, companies from Louisville, Cincinnati and Dayton, six abreast, wearing white scarfs with black rosettes. Next come the students from the Western Military Institute, dressed in a splendid uniform of white, each wearing a long black scarf trimmed with a white rosette on the shoulder, all being armed according to the Military Code of the United States. They look particularly sad as they slowly march along with arms reversed, and seem to attract considerable attention.

The corpse is followed by the company from Frankfort, the Capitol, one of the oldest in the State, and together with various other companies from places not known to me they make a long line, and quite as fine a display as any thus far in the procession. Following this military array is a fine band of music, discoursing solemn and touching strains which seem to send a thrill of sorrow throughout the whole city.

Next, carriages, two abreast in an almost end-

25

less line, conducted by their Marshalls. Following these in close order are the members of the 'Grand Lodge' of Kentucky (Masonic), dressed in the regalia of their Order, all in deep mourning, and by their sad looks seeming to realize that they have lost a worthy brother and the brightest ornament of their Order.

Immediately in the rear came the remains of the mighty Dead, in a style of sad magnificence which almost beggars description. (Here you must suffer me to digress from my notes for the purpose of giving you a description of the 'Funeral Car' from memory, as the time it was in passing did not allow me to do it at the time, and indeed I felt a sadness come over me as the remains of this glorious old chieftain were passing that entirely unfitted me for writing. I could not, at the moment command my self sufficiently to put my thoughts on paper, nor did I try, but sat gazing upon the scene before me until other things caught my attention.)

'The Funeral Car' was drawn by eight pure white horses, each entirely enveloped in drapery, trimmed with silver, and attached one to the other by silvered chains which were placed upon either side. and thus passing the entire length of the team. The horses were hitched to the Car two abreast, and were led by eight negro men, one walking at the head of each horse. These negroes were dressed in white, with black hats around which were tied long white crepe scarfs, which descended to the knees, and upon their arms they wore black badges. The Car itself was fitted up especially for the occasion, and at great expense, and was surmounted by the American Eagle in a hovering position, holding in his beak the four corners of the drapery which enveloped the entire Car. The Eagle was perched

26

upon a vast silver urn, from which he seemed to look down with sorrow upon the scene below. I understood from one who knows that the extra trimming of this Car cost $800, and I do not in the least doubt it, for the inside was lined throughout with fine satin, and everything else was in proportion. Now by your permission we will return to the notes.

Immediately behind the Car were the Congressional Committee and Pallbearers, in carriages, followed by a long line of citizens and strangers in carriages two abreast, some of whom had their horses trimmed with mourning. Following these were the 'Odd Fellows', with all the insignia of their Order - they don't look quite so impressive as the 'Masons'. Next came the various fire companies of Lexington, Louisville, Cincinnati and Frankfort, marching in silence, each dressed in so much taste that it is difficult to make a distinction between them.

And finally the procession was closed by citizens and strangers on horseback, four abreast, not by any means in uniform, but each one dressed according to his own sense of propriety, and all seemingly impressed with the solemnity of the occasion and keeping the most perfect order.

At the cemetery gate the whole procession is forming upon foot and is slowly wending its way, by circuitous paths, to the final resting place of the Dead, - the spot especially chosen by Mr. Clay for his dust to sleep.

CHAPTER FOUR

Iowa

Jennett, Lyman, Caro, Baby Girl
leave the nigger Mammy and Kentucky

IN eighteen hundred and fifty, the Bennetts had moved from western New York to Rock Prairie, Wisconsin. Lyman and Jennett were anxious to join them, and began early to plan on homes nearer each other. Lyman, by this time, had decided to quit teaching and go on to a farm; and he felt that Kentucky was no place for him. The parents of Mary Lizzie felt strongly the urge of getting over the 'Mason and Dixon' line; 'Equal rights for all' was born and bred in them. Jennett wrote to her parents when preparing to leave Kentucky, "If I had not the best neighbors in the world, I know I should have the 'hippo' all the time, but they all seem like mothers and sisters to me. There are so many kindly attentions. There is nothing they have, that they know I have not, but they will divide, and it grieves me sorely to leave them."

Both Lyman and Jennett had grown tired of teaching, and tired of living among aliens, for such they must regard their Kentucky friends. Dear as were these friends and their sunny home, they knew the dividing line was there and must be crossed. So they sold the Kentucky home and all the household effects. Lyman expected to take his family to Rock Prairie in March or April of eighteen hundred and fifty-two, but it was a year later, in

March of eighteen hundred and fifty-three, before the move was made.

When Mary Lizzie was not yet three months old she was taken from the arms of her black mammy into the arms of her own mother and bundled into the stage along with father and sister Caro. At Louisville a lake steamer received the pioneers -- Kentucky was behind them.

The lake was stormy and the voyage across rough. When the night was at its darkest there sounded the dread call, "All on deck for the life boats." Wakened from sleep the poor mother threw on her clothes and reached for her baby - but there was no baby! Frantically she searched the berth, the cabin, the salon, and the deck. No one had her, no one had seen her! Back into the cabin a terrified mother rushed, and down to search again the scant space under the berth. There at the back Mary Lizzie lay curled up asleep. Some time in the night the pitching of the ship had thrown her to the floor. Like the good little scout that she was, she had rolled about hunting warmth until she found it, and there had curled up and gone to sleep.

By the time she was found, the worst of the storm was over and the passengers were ordered back to their cabins. Eventually they were landed at the port of Chicago. Again the stage received them and carried the little family of four to the haven of Grandfather Bennett's in Rock Prairie in Wisconsin.

A letter written by Frank Nash, son of Lyman's sister Catherine Curtis Nash, gives an interesting account of the gathering of the clans of Bennetts, Nashes and Curtises in Rock Prairie. The Bennetts were first, moving west from New York state in 1849 or 1850. The Nashes followed in 1851. The first of the Curtises, William Van Bergen Curtis and his wife Salina, arrived in the winter of 1851-52, and were domiciled with the Nashes until their own house was ready.

A PIONEER GIRL

The Bennetts had a frame house consisting of living room and kitchen, with pantry cut off from the bedroom beyond the kitchen, and over all was a good sized loft. In this small house lived Jacob and his wife Caroline together with the five remaining children, William, Lewis, Corty, Libby and Vina; and yet there was room and a genuine welcome for Lyman and his family of four.

Lyman and Jennett had expected to locate here, and they did remain through the spring and summer months, while Lyman helped to seed the fields and harvest the grain; but Iowa, with her unbroken prairies, was calling for settlers and the government was selling her land at a dollar and a quarter an acre. Lyman saw his opportunity there.

In the fall of eighteen hundred and fifty-three he went on into the Iowa wilderness to prepare a home for his little family, leaving Jennett, with Caro and Mary Lizzie, safe in her father's care. The anxiety of that first winter, and the loneliness of the Iowa forest formed a sombre background for the picture of Mary Lizzie, who rolled on the floor and kicked and gurgled and crowed her way into all their hearts. Lyman's two young sisters, Elinor and Mary, were a part of the Catherine Nash family that winter, and chums of the Bennett girls, Libbie and Vina. Happy Mary Lizzie had only to hold out engaging arms to be caught up and carried where the little pigs rolled and grunted. To the merry girls who carried her it was a never failing source of delight when she squealed and grunted back to the squirming mass and held out her arms for them. She knew nothing of her father alone in the woods, nor of the Indians on the warpath. That year Minnesota Indians were reported as following the forest down into Iowa, leaving everywhere a trail of death and desolation.

Father Lyman was not to be held back by rumors; his plans were made, and he went ahead with them. How steadily and forcefully he worked through these fall and winter days, the well-made cabin of logs squared and joined with New England exactness told better than words. The oaken floor was smoothed with infinite care and the trundle bed, stools and chairs were all worked down with plane and draw to splinter-free smoothness. The dining table of native walnut was a work of art. Its six legs were turned, bound and finished with brass rollers, the wide side leaves had smooth roll joinings and were hung on brass hinges -- Lyman's work, the whole so carefully done that after three-quarters of a century it still rolls a perfect table.

Lyman had built the cabin in an open spot among the oaks and hickories, within easy reach of a fine spring that was the crowning jewel of his timber forty. Close by the cabin a small garden had been wrested from blackberries and wild roses. The seeds were planted and the whole protected by a high worm fence, each corner stayed by stakes that carried a rider from crossing to crossing to crossing. This high fence was necessary to keep out the deer. Timber was plentiful, and the wild life of the forest, all unafraid, was curious and hungry, and must be kept out of the garden and grain fields.

Lyman paid a dollar and a quarter per acre for his land; one hundred and sixty acres of prairie and forty of timber. The prairie land was in Bremer County and faced east, lying along the road that separated Bremer County from Fayette County; Because of this fact it was known as the 'line' road. A mile of land (not yet filed on) lay between the one hundred and sixty acres of prairie land and the forty acres of timber, which was part of the woods bordering the Wapsipinicon River on the east, and formed the boundary to that great prairie that outstretched for miles to the west. Here Father Lyman later on erected the frame house that was to be their permanent home, and ten acres, fronting on the county line road, was set aside for house, garden and farm buildings.

It was late in the spring of eighteen hundred and fifty-four before the home was ready for Jennett and the children; the birds were nesting and the wild wood flowers were all out to welcome them when Lyman brought them to the new home. Wild plums were in bloom, grapes rioted over everything; the oaks and hickory-nut trees, standing in sturdy strength about the cabin clearing, were fringed with blossoms; the hazel bushes, clustered at the edge, were full of catkins; everywhere there was promise of plenty.

Ambrose McKee and his wife Margaret were their helpers. They were typical Scotch people who had come into Iowa about the same time Lyman had, but unlike him, they had not entered any land (they bought later a piece already proved upon); both were strong and healthy, both were good workers. Land was plentiful and could wait, while board and wages could be had at Lyman's.

With the help of Ambrose McKee and his oxen, fields for oats, sod-corn and pumpkins were broken up, dragged, and planted before Lyman could leave to bring Jennett and the children home. Once the seed was in rich black loam, Nature would take care of it while he was gone. Diligently as he had worked, it was late in the spring before everything was in readiness; he turned loose the cow and her young calf to forage for themselves, locked the cabin door, climbed into the lumber wagon, gathered up the reins,

and started on his long trip to Dubuque; twenty-five miles to Independence, where he stayed the first night, and seventy to Dubuque, but the big horses, Dave and Jack, trotted off as though the lumber wagon was a plaything; and the enforced rest was good for Lyman's tired muscles.

At Dubuque he left his team, and steam cars carried him on. Already in fifty-four the railroads were reaching west from Chicago, and time was precious. For this same reason Grandfather Bennett brought Jennett and the little ones into Chicago to meet Lyman, and the reunited family were soon on the homeward way.

Just ahead of them tragedy grim and ugly had crossed and left terrible wreckage. Mary Lizzie saw a horrible sight from the safe shelter of her mother's arms. Two rival trains, racing west, had claimed each the right of way to a common crossing. Both had crowded on steam and raced ahead through the darkness, each striving to be first. Neither would yield, but tore ahead for the terrible crossing. There followed the crash, screams, the hiss of steam and the mounting flames. It all formed for a little space a part of Mary Lizzie's life. Travelers from the following train were forced to leave their cars and pick a devious way through the pandemonium. Once they were across, a train from the west met them and carried them on.

When Dubuque was reached, Mary Lizzie was very sick, and for anxious days they fought for her life. Dr. Langworthy and his wife, after the generous habit of these Westerners, took the little family right into their home, where the skill of the physician could be backed by the safety of home comfort. Lyman could not wait; again he must travel alone the long road to the prairie farm in Bremer County, but he knew his family was in safe hands. He packed into the big wagon things that were necessary; a rocking chair for Jennett, a looking glass, a bed-stead, whose four high twisted posts and the head-board and foot-board of twisted rounds were all of polished cherry. Each side-piece was fitted with the little knobs around which to lace the bed-cord, that formed a safe bottom on which to rest the tick, filled only with clean oat straw. A big feather bed was to be placed on top of that. A few iron cooking utensils, some tin ware and dishes completed the modest outfit, bought in Dubuque. A walnut chest of drawers, Lyman's work, was waiting in the cabin where a cook-stove had already been installed. The trundle-bed frame with slats stood in place ready for the mattress and covers, before being pushed under the big bed.

Some weeks later Lyman, in the lumber wagon behind Jack and Dave, traveled again the long road to Dubuque. It was a thankful father and mother who finally packed their little stock of necessities into the big lumber wagon, lifted Caro to the seat between them, and with Mary Lizzie, again a well baby, cradled in her mother's arms, started on the home trip.

From Dubuque to Coffins Grove the first night, on to Independence for the second night, then a long day's drive of twenty-five miles more to the home that waited at the end of the road, literally the end of the road, for beyond that lay timber and unbroken prairie sod.

Letters from Lyman to Jennett's brother Nehemiah, written at this time from Coffins Grove and Independence, in the summer of 1854, show that already he had become a leader, for these letters were arranging for proper filings on land grants for distant neighbors. Other letters written later in the pioneer years show him to be a recognized authority on questions of law - "pettifogging" he called it. There were so many who needed help, so often an injustice had to be righted, that Lyman had many calls on his knowledge of Law, and needs must study the code to keep abreast.

Jennett wrote: "Once in a while when Lyman gets a little money from his office, (viz.) Justice of the Peace, we get a little sugar and other groceries. Lyman is also Town Clerk and School Director." Farther along in this same letter she referred to Lyman's law work as follows: "Lyman got through with one case yesterday, and has gone on another lawsuit. I do not know but he will turn lawyer in good earnest. They seem determined to keep him busy."

Here Lyman took up the pen: "Coming into our parlor, which, by the way, you will please to understand is not finished, I find Jennett's letter on which, without asking to know if it is finished on her part, I take the pen and scribble away. I perceive that she has been telling you some very important and I presume to you interesting news in relation to my business. It is true that I do sometimes act in the capacity of...of...of...why, Pettifogger, I suppose is the term to be employed in the case. Now I hope you will conclude that it is not my intention to become a professional disciple of Blackstone as Knight of the Green Bay. Yet when I see one likely to suffer injustice for want of legal advice and assistance, I own that I do appear in said very disreputable capacity for a fee, and this last I am sure you will agree with me, has car -

ried comfort to the heart and brightened the face of many
a better man than your humble and obedient servant who
is now addressing you - then do not think too hardly of me
(for the time being) for disgracing myself in company
with such a set of scoundrels as lawyers are by some
thought to be."

Wherever Lyman lived or among whatever people his
lot was cast, he was a leader. Men acknowledged his
ability and trusted his stern integrity. He belonged to the
highest type of men that New England ever sent out. His
wife belonged in this same grade. The East never sent
forth any couple that were finer or better than these two.

The Cabin at the end of the Road

CHAPTER FIVE

The Forest Years

The Forest Years

In the forest, heaven is near on a summer's day
The scent of the flowers, the hum of the bees,
Comes drowsily sweet 'neath the forest trees.
See how they nod and beckon and sway
Oh, heaven is near on a summer's day.

L.C.F.

IN the woods that bordered the sluggish black water of the Wapsipinicon River, Lyman had built the cabin for his family. Here was home, and here Mary Lizzie thrived and grew from the home made walnut cradle and the high-chair, to be a toddler, always under foot, caught up in someone's arms to be danced and put back in the high-chair by table or window. In at the open door, a pet fawn raised by one of the distant woodsmen, a Mr. Ship who lived eight miles below, came to be a playfellow, eating turn and turn about from the A.B.C. pewter plate or nibbling from the crust in chubby hands. Between the pink nose of the fawn and the red mouth of the baby, many a mug of milk was spilled onto the floor - a platter de luxe to pink nose, shared to the best of her ability by little red mouth.

Squirrels scampered along the top rail of the fence and

onto the window ledge, in and around Mary Lizzie whose arms were kept busy warding off the tickle of a bushy tail. Jack, the big black horse, on his way to the log stable, would put his nozzle in at the window to nod at her, waiting before he followed his mate to the manger, for the welcoming clap of little hands or, if Mother Jennett had time to lift her up, for the kiss to be put by baby lips on the white star on his forehead.

When winter snow was deep over everything, from her high-chair inside the window, Mary Lizzie's big eyes saw the stately deer lead his flock along the fence to lick the snow from the top rail, where a handful of salt or a few kernels of corn or some grains of wheat from Father Lyman's treasured food supply had been scattered.

The deer who came to the fence for
salt, grain or something to eat.

Father Lyman had endless tales to tell at night of the meadow larks that were so numerous, of the striped snake that got himself tangled up with the breaking plough and rode to safety along the beam, of the pocket gopher that pocketed her babies and scurried off when her home was disturbed, of nests of field mice whose swinging houses went down before the ever advancing plough, and of the mother quail, whose nest was rescued from under Jack's big feet and carried to a place of safety where she again covered the speckled eggs and finally hatched out a big brood of chicks. Once he told how Ambrose McKee, his axe over his shoulder, swinging along through the tall grass of the slough to the west, was followed all the way by a big timber wolf. When the brute came too close Ambrose would turn, swing his axe and yell till the wolf would skulk to the rear. He was never far behind until the old Scotsman climbed the fence that was Lyman's western boundary, where the wolf sat down and watched him clear across the field. After the stories came the trundle bed, pulled out from under the big bed, where two sleepy little girls were tucked in by Mother Jennett.

These were happy days for Mary Lizzie. She was steadily growing from babyhood into the little girl who romped unafraid with sister Caro around the cabin in the woods. She knew nothing of the savages rampaging down from Minnesota through the forests of the Wapsipinicon River, leaving everywhere their bloody trail. She knew nothing of the fear that clutched Father Lyman's heart

when he tossed her up in the morning for her goodbye
kiss, lest at night time he might find only ashes; knew
nothing of the agony of fear with which Mother Jennett
locked her babies in the cabin, and ran with her two pails
to the spring for the water which she must have, dreading
each moment lest she hear the awful war-whoop.

For Mary Lizzie these years in the forest were filling
her being with a love of nature that never left her. She
neither knew nor cared that the fine linens of her Kentucky babyhood had been replaced by soft muslins and
figured prints or calicoes. They were always pretty, for
Mother Jennett was the same artist here that she had
been in the happy southern days. The white curtains at
her cabin windows, the rugs on her floor, the chintz that
cushioned her one rocker, and the grandmother spread
that covered the big four-post bed in the corner were all
artistic. Jennett was fortunate in possessing two of those
wonderful spreads woven for her by the grandmothers.
One was in soft old pinks, the other of grey-blue that
shaded into indigo. They were both the product of the
hand loom, and were blocked off in small squares. Many
were the steps taken by the weaver, and ceaseless the
beat of the loom bar to bring forth the firm product. Before the weaving had come the carding and washing of the
wool, the spinning of the yarn, the coloring and the intricate threading of the loom. Mary Lizzie's grandmothers were possessed of artistic souls, deft fingers, and
infinite patience. Mother Jennett was never a weaver.
She had neither loom nor spinning-wheel, but both Grandmother Bennett and Grandmother Curtis had kept their
spinningwheels and looms busy, and turned out the household linen as well as blankets and the soft material that
clothed their families.

Meanwhile the story of the rich black soil of the Iowa
prairies, with Government land at a dollar and a quarter
per acre, was abroad and settlers were crowding in. It

was the time of doggerel verse and catchy song. From the wagon train floated back such songs as:

> Jump on the wagon,
> Jump on the wagon,
> Jump on the wagon
> And we'll all take a ride.
>
> Come along, come along,
> Don't be alarmed
> For Uncle Sam is rich enough
> To give us all a farm.
>
> Come along, come along,
> Don't be a fool
> For Uncle Sam is able
> To send us all to school.

The land fever was catching, and it spread fast and far. Those first years of life in the forest were filled with the tramp of the land seeker, and the few cabins of

the early settlers were long distance mile posts for the
oncoming flood of pioneers. Time and again the cabin loft
would be filled to over-flowing with men who sought shelter from the weather. Time and again, when the loft was
full, they brought their wagon blankets and slept on the
floor. Father Lyman would have to rout them out in the
morning before Mother Jennett could get up. Everyone
who sought shelter found it and a welcome in the cabin in
the woods. In these times the walnut table was crowded
by a hungry set of men; and in self defense Lyman and
Jennett found it necessary to make a small charge when
men were able to pay. Father Lyman built a lean-to along
one side of the cabin and here kept for sale such staples
as were necessary to pioneer life. These were more often exchanged than sold, for money was not to be had.
Life was carried on through the medium of exchange; men
offered what they had and took in return of the others'
surplus. Fish, prairie-chicken, grouse, and even an
occasional wild turkey were exchanged for sugar, salt,
flour or meal. Venison was always a luxury, but many a
hind-quarter was pegged to the wall where it would be
safe for use when wanted. A well tanned deer hide made
a comfortable resting place for bare feet.

A neighboring mill, Chitister's Mill, the next mile
post on the north, three miles away, sawed up the forest
logs into scantling, sheathing, rafters, shingles, siding
and flooring. Across the big prairie ten or twelve miles
to the west, Buck Creek Mill ground up the corn and
made the wheat, rye, and buck-wheat into flour. Everything that could be used went into building homes and into
keeping life in the bodies of the settlers.

The cold, the bleakness, the tragedies of life had no
terror for Mary Lizzie. Her days were warm, her life
was a merry one. If she heard the long drawn howl of the
big timber wolf, or the short yip-yip of the prairie wolf,

she and Caro fled to the cabin and their mother. Her days were full from morning until night. She was lifted up to stroke pityingly the slaughtered deer in the passing wagon, whose owner left meat in exchange for sugar or salt. Father Lyman was not a killer, but he took in exchange; and the deer, the squirrels, the grouse and quail lived their lives unafraid of the cabin in their midst. He never owned a gun during those pioneer years when life was lived so near to the danger line. Yet Mary Lizzie once heard her Uncle Will tell of a wager as to who was the best marksman in the group, and Lyman carried off the honors. His eye was absolutely true and his nerve unflinching. The unwritten law of life for both Lyman and Jennett was honor. Men believed in their truth and knew that their dealings would be just.

These first hunters were not men of blood lust; they killed that children might eat. Mary Lizzie was shielded from the knowledge that her missing playfellow, grown into a splendid buck, met the common fate.

An occasional panther came to the neighborhood of the cabin, and his blood-curdling scream carried terror even to the brave heart of Mother Jennett. For days she kept the little ones in the shelter of the cabin, and went only to the spring when dire necessity compelled.

The children of Dutch John, one of the distant forest neighbors, were followed by a big timber wolf clear to the cabin door, where the Dutch woman beat it off with her broom and finally drove it away. The big timber wolf was always to be feared, and when the snow was deep and the cold intense, even the cowardly prairie wolves became ravenous and were a real menace to children. At such times, when a pack of wolves was on the hunt, cattle gathered in a ring with calves in the center, and fronted the danger with lowered horns. Even ravenous wolves dared not break through that circle. But woe to the wanderer when hunger drove the wolves! Only an occasional bear was to be found in the Iowa woods, but the fear of them was ever present.

. When wee brother Willie was born and filled Mother Jennett's arms, it was to the kindly Scotch woman, Margaret McKee, Jennett's helper, that Mary Lizzie went for mothering. When bread and butter fell on the ground, it was Margaret who wiped away tears and brushed off dirt with her soothing, "There, there, 'Little Mother', don't cry, it didn't loss nothing." It didn't, and 'Little Mother' was well content to finish her lunch happily.

A time came when the sturdy Scotch blood of Ambrose McKee rebelled over hiring out to another man, so he secured a little place to the west of the Curtis farms across a corner of that great prairie. Lyman and Jennett had to turn elsewhere for their help; but their places were never filled in Caro's and Mary Lizzie's hearts.

It happened once that Jennett yielded to the pleadings of the two little girls and let them go all by themselves for a visit with Margaret McKee. Jennett had perfect confidence that the Scotch woman would see them safely started on their homeward way in good season. Margaret did this and Mary Lizzie and Caro, after a happy visit, started for home, skipping blithely along until, from behind a knoll they had just passed, a prairie wolf came out to watch them. No more larking along. Two little girls holding tight to each other's hands sped over the grass just as fast as their legs would carry them. The wolf kept an even distance to the rear. It was quite possible that the wolf, never before having seen such a whirlwind of waving sunbonnets, swelling skirts and flying legs, nor heard such an outcry, was half afraid himself. However, he made no demonstration against the children but kept his even distance behind, through, or over or under the boundary fences, across the ploughed fields, under the high stake and rider fence that enclosed the ten acre lot, where the house, barns and stacks were - clear up to the door-yard, where Mother Jennett received the little girls and added her compelling voice to the general outcry. The wolf slunk away into the shadows of the tall slough grass. Father Lyman, working in a distant field, knew nothing about it until he came home at night. Never again did Mary Lizzie and Caro beg to cross that big prairie alone.

CHAPTER SIX

The Forest Years Continued

THEY were so comfortable in that cabin in the Wapsipinicon woods that the forest years stretched out. The Indian outbreak had been quelled, at least for the present, and Lyman needed all his time for getting the prairie acres fenced and cultivated.

There was beauty all around them, wild roses, lush buttercups, graceful sprays of white anemones. The beds of violets and spring beauties, that bloomed in the spring, gave place to the tall spikes of the cardinal flower, the wax of the mandrake, and the riot of plums and cherries, wild crabs and blackberries. These were followed by the crimson of the sumacs, red and black haws, hazelnuts, hickory nuts, walnuts and butternuts, all set in a background of purple grapes and red oaks.

Another touch of cold and the beauty was gone. Dead leaves went scurrying down to cover the earth. The deeper cold sent the acorns rattling down, some still in their mossy cups; walnuts dropped their shiny green globes to the ground; the butternuts hung on a little longer, while they hardened the spikey burrs that grew underneath to clutch fast the green overcoat. The hickorys were best of all, for their paneled coats unfolded and the glistening

46

white nuts pattered down on the bed of leaves and moss for Mary Lizzie and Caro to pick up. Along the edge of the forest and in the sunny spaces were the hazel bushes, their clustered nuts within easy reach of the children. The grape grew sweeter, the blackberry more luscious, and the wild plum deepened its blotched golden pinkness. All this wild growing, glowing beauty cast a spell over Mary Lizzie, who loved every bud and blossom and fruit.

There were interludes in the forest existence. Once Grandfather and Grandmother Bennett and Jennet's young sister, Vina, came to brighten life in the cabin through the summer months. Their visit was a round of picnics and exploring walks, and Mary Lizzie's knowledge of the forest was broadened. Holding on to the hand of her beloved Aunt Vina, she danced through the forest and saw in every glade something that was new and fascinating. Once a deer, frightened out of its mid-day nap, in one graceful leap was up and over the young tree in whose shade he was resting; and once a mother doe led her baby down the path before them. Sometimes a mysterious drumming came from some place ahead, and Mary Lizzie stopped suddenly to listen, half afraid until she knew it was only a partridge, and once they stole so quietly along that she had a glimpse of the little gray drummer circling the trunk of a big oak tree.

Life for the Curtises was never monotonous. Through it all Mother Jennett moved busy and contented. Her letters were just a cheerful chronical of every day life, all closely written, rather bulky, the outside page left clear for the folding and address.

The folding was intricate, and when it was finished the pointed flap was turned down and sealed with its drop of wax. Along the upper right hand corner of the address side was occasionally written a guarantee that twenty five cents had been paid for the safe delivery of the letter.

Thankfully the father and mother did hard, bone-aching

work to wrest this home from the wilderness. These pioneer years were happy ones. The cabin had become more than a temporary shelter. It was home for the Curtises.

In winter the snow swirled down and settled on trees and bushes, until the slender branches bent or snapped under the strain. The heavy white coat packed in smooth open stretches was a warm blanket for Mother Nature to tuck around her babies. Inside the cabin there was a roaring fire with nuts to crack - hickory nuts, hazel nuts, walnuts and butternuts. Nine o'clock was bed-time for everyone. The children went earlier. They undressed and put on their night gowns in the warm living room. Night-caps were tied, prayers were said, and clothes were gathered up; then followed a quick little run across the cold bedroom floor and a jump into bed, which Mother Jennett had opened. Then they cuddled down into the depths. The warm bed clothes were carefully tucked in place before Mother Jennett bent to put her goodnight kiss on each forehead. Out went the light, for candles were precious and little eyes were already shut.

In the fall of eighteen hundred and fifty-four, Newman Curtis, Lyman's father, accompanied by Henry, one of his younger sons, had come into Iowa to look the land situation over. Later he moved his family out.

Grandfather Newman Curtis was a good financier. He not only supported and educated his big family, but when he arrived here in Iowa he also had twenty-five thousand dollars. Four thousand dollars of this he carried in twenty dollar gold pieces in a buckskin belt worn next to his skin. After he had rented the Daniels farm for one year, he bought two hundred and forty acres two miles north of Independence, from a Mr. Harwood, at twelve dollars and a half per acre. H. P. Rozell made out the Deed, and the two men were astounded when Grandfather Curtis took off his belt and began counting out twenty dol-

lar gold pieces to pay for the land. They had never seen
so much money, and Mr. Harwood said he could not tell
whether it was gold or brass; and anyhow he had no way
of taking care of so much money. He added: "Here's your
Deed. Take the money back to Dubuque and deposit it
in Bissel and Company's Bank." Grandfather's skin was
raw from wearing that belt, and he offered Mr. Harwood
twenty dollars more if he would take the money, but the
reply was, "No, not if it were one hundred dollars." So
it was up to grandfather to carry the three thousand dollars back to Dubuque and deposit it in the bank. This oc-
curred in the days when banks were few and far between.
Grandfather's farm was still father away from the bank
than was the little town of Independence. Money was kept
in the house. Dubuque was more than seventy miles away,
and the only means of transportation was by horse, but
Grandfather Curtis made the trip and deposited the gold
in Bissel and Company's Bank.

Wilbur, his youngest son, was eleven years old when
they entered Iowa, and the next year his father took him
back to New York to an older brother, Dr. John Curtis,
and left him to go to school. The poor boy was so homesick in his letters that his mother wrote him he might go
to an old friend whom Grandfather had helped to get a
start, and borrow the forty dollars necessary to bring
him home. She thought a visit would cure his homesick-
ness, and then he could go back. He borrowed Dr. John's
horse and buggy on the plea of going to see the old place,
and drove over to the home of the English friend. This
was in eighteen hundred and fifty-seven, remember, when
banks were not so common. The old Englishman kept
his money in his house. He was haying when the boy arrived, and up in the haymow, but when he found that the
boy was Wilbur, a son of Newman Curtis, he came down
and said, "Wife, go to the safe and get forty dollars. I'll
let Newman Curtis's boys have money if they want it."

Better still these good people arranged to see Wilbur
safely from Medina, New York, by stage, into Chicago
and onto the train for Galena, Illinois. At Galena he would
again take the stage to Independence.

He made the trip in safety, and it was a happy boy that
walked the two miles out to the Curtis farm. Mother
Maria welcomed him warmly, but when Father Newman
arrived an hour later, Wilbur described that time as "the
darkest spot in my life." Grandmother Maria usually had
her way, but this time Grandfather Newman decreed that
Wilbur had had his chance, now he could go to work on the
farm. At that time there was no school other than a private school for young men in Independence, and this was
but little more than a school for beginners; in these early
settlements, schools were not yet developed; but they
were not long in coming.

Willie was born in the cabin in the woods early in the
fall of eighteen hundred and fifty-five. Not long after-
ward, in the early summer of eighteen hundred and fifty-
six, hoping to benefit Jennett and the baby, who seemed
frail, Lyman took her and the three children to Dubuque
and put them aboard the river steamer bound for Red
Wing and her father's new home, the Tepa Tonka Hotel.
(After Lyman and Jennett had gone to Iowa, the Bennett
family was not long in following their two boys to Dubuque,
where they remained for a few years and where Elizabeth -
or Libbie - met the man she afterwards married, Stephen
Langworthy, son of the Dr. Langworthy to whom Baby
Girl owed her life. Eventually the whole Bennett family
except Lyman and Jennett went up the Mississippi to the
little town of Red Wing, Minnesota. Grandfather Bennett
bought the Tepa Tonka Hotel, and here Aunt Libbie was
married and the younger children were educated in the
State Normal located at Red Wing.)

Again Lyman returned to the sombre loneliness of the

great woods, content that his family would be well cared
for. There was plenty of work for him early and late to
keep him from feeling too keenly the emptiness of the
cabin room.

Of their Red Wing visit Mary Lizzie has only a few
cloudy memories, and these are of her play with a little
cousin, Uncle Neal's little daughter Nettie; but there
stands out sharp and clear the 'pop' that came when a
bottle of ginger ale exploded in her hands. This is all
that memory can show until the time came for their return, but she knows that for a space wee brother Willie
nestled in her mother's arms; then she remembers the
tiny coffin they carried onto the steamer, her weeping
mother and a bird cage, with a little nankeen cape pinned
over its top, wherein was a sweet voiced canary put into
her hands by the beloved Aunt Vina, to soften the grief of
the parting.

Father Lyman met his family in Dubuque, and the little coffin and the canary bird in its cage went with them
in the lumber wagon to the lonely home in the woods. The
baby was laid to sleep out on the prairie farm under the
sunshine and the starlight. For Mary Lizzie there was
no sadness in this home coming. She was not old enough
to understand what had happened, and was just glad to
nestle once more in her father's arms and to see the big
horses, Jack and Dave, the cows, the little pigs in their
pen, the kittens and all the things that a child loves.

In the forest cabin, Mary Lizzie was growing up. Her
straight black hair went under a plain cap at night, and
in the daytime two braids held it out of her eyes. The
sunbonnet tied on by mother's careful fingers was soon
danced off to hang where Mary Lizzie evidently believed
it belonged, down her back, while the heavy black braids
streamed out behind. She romped with sister Caro in
the sunlit space where the cabin stood. Together they
swung in their grape-vine swing, made secure by Lyman's

51.

careful hands -- swung, not very high, but quite as high as was safe for little girls to swing.

For a playhouse the two little girls used the smooth flat top of a big granite rock, rising a few feet out of the ground. Nature had fashioned a stairway of broken stony steps, easy to climb, and if a tumble happened in the climbing, the thick moss at the foot made a soft bed on which to roll. Once up, the rough surface of the top prevented slipping, and there were never any serious accidents. Acorn dishes were not breakable; neither were the rag dolls. There was nothing to catch toes under, and snakes could not reach them. Here was Mary Lizzie's and Caro's favorite place for afternoon tea, served to the dolls out of mossy acorn cups. Mother Jennett added as the finishing touch to their rock playhouse, a thick, soft old comforter that made for the two little girls and their beloved rag dolls a wonderful resting place when the afternoon sun streamed in over the surrounding thickets of wild plum and hazel. Overhead the squirrels chattered, and scampered along the branch of the big oak that shaded the rock. The robins and the orioles joined in the general visiting, and even the shy wren peeped unafraid from her nest to see what was going on.

No playhouse could have been finer. That big granite
rock was one of the treasures Nature gave to the little
girl. Long after she had grown up and gone, her beloved
rock was blown out and went to take its place in the foundation walls of Iowa's State Capitol. Perhaps it had
missed the patter of little feet over its level top; perhaps
it was glad to have its pieces cut and polished into things
of beauty, and placed in those foundation walls to be forever of use among its own kind. So the little girl-grown
woman who touched tenderly the gleaming black and white
of those polished surfaces likes to think.

Much of the finer work for the prairie house and barn
was done evenings in the cabin, by firelight and candle-
light. Lyman's work bench stood under one of the kitchen
windows with the big tool chest at its end, and Lyman
with plane and adze, with saw and hammer, worked steadily throughout his spare moments. Frames for doors in
the prairie house were set up, and glass puttied into the
window casings. Flooring was planed, and shingles made
ready for the roof. Father Lyman's plane rolled off
beautiful long even curls, and it was an endless delight to
Mary Lizzie, with her straight black hair, to cover her
head with them. At four years it was easy to imagine
herself a real curly head.

It was at this time that he made the fine old secretary
of black walnut lumber, all so well seasoned; the joining
so perfectly done, with never a nail to mar its surface,
that now, three quarters of a century later, after traveling over a good share of the western part of these United
States, it stands in the New Mexico home with never a curl
or crack or drawn joining, fair as when it left the maker's
hands. No varnish ever stained its fine old surface, but
years of oiling and rubbing have kept it free from scratch-
es. The legs were turned, like those of the table, from
the same Iowa walnut, that there might be no failure in

perfectly matching the rest of the frame work. The inside
of the secretary was fashioned to suit Lyman's needs. At
the bottom was a drawer for stamps, pencils, pens, etc.,
and over it were the spaces for long ledgers; pigeon-holes
for various sorts of envelopes were to the left; above was
a broad space for paper, and over that the shelf for the
family Bible; then rows of pigeon-holes all carefully
labeled, while at the top was a full length shelf for mis-
cellany. The two doors met in perfect accord in the center
of the case, which stood on a table with a writing leaf that
could be raised or lowered and had under it a space for
keeping account books conveniently at hand.

While Lyman worked, Jennett's busy fingers kept the
knitting needles clicking merrily, whenever the thimble
and needle were laid aside, all by candle light, and yet
both Lyman and Jennett had at eighty, keen, bright, old
eyes that read with spectacles pushed up out of the way.
Was the softer light of the candles better for the sight
than our own electric brilliance? I wonder!

During the winter the fence-posts and rails were cut
and strung along the line, ready to set up when spring
came. The stone for foundations was hauled, and logs
were cut and taken to the mill to be cut into lumber that
should frame, sheathe and shingle the house and barn.
There was to be a big barn with a runway for pigs underneath, with great bins for wheat and oats, a big hayloft
and storage room to hold over farm products until such
time as Lyman wished to sell. He planned well, and the
time of need for which he prepared actually came.

During the winter of eighteen hundred and fifty-seven
and fifty-eight the bench work for the house and barn was
finished, and as early as possible in the spring of fifty-
eight, the frames went up and the buildings were completed. Neighbors came from far and near to help in
those raisings. Later on Lyman and Jennett repaid their
kindness by throwing open the big new barn for a Fourth

of July celebration, to which the whole country side was invited, and duly responded; the families coming in lumber wagons from miles around, with numerous babies and children, and lunch baskets filled to the brim with all the good things of that early day. Mary Lizzie remembers one very special treat: some little scalloped cakes of maple sugar. Candy or any such sweet was indeed a rare treat in those pioneer days.

The regulation Fourth of July celebration, with the reading of the Declaration of Independence, and a patriotic speech by someone well versed in such oratory, was followed by the big picnic dinner. This was a gala occasion, one of the few occasions when those early settlers could get together, and therefore more enjoyed and longer remembered.

When once the prairie home was ready, the moving and settling had been quick work. The new rag carpets which Jennett had ready were laid in the living room and bedrooms, the stoves went in, the curtains were put at the windows, and the big walnut table took its place in the center of the living room. A comfortable new chair for Father Lyman's desk was added, the few precious pictures were hung, and they were ready to begin life in the new home. Wherever Jennett lived, she made life attractive.

The pictures that hung in that living room were worthy of note. One, the work of Jennett's brother, Nehemiah, consisted of 'The Lord's Prayer' done in the flowering of Spencerian penmanship from the beginning of "Our Father" to the "Amen." It was one series of vines, scrolls, curlicues and flourishes, all perfectly executed, and was set in a walnut frame. It had the place of honor between the windows in the front wall, and was flanked by Benjamin Franklin and George Washington. Around the walls hung other presidents and men of note - John Quincy Adams, Thomas Jefferson, and Henry Clay, along with

William Henry Harrison.   With the scraps of doggerel
used in the presidential campaigns Mary Lizzie was familiar, and liked their jingley sound, although they had
little meaning for her:

> With Tippecanoe and Tyler too
> We'll beat little Van
> Van, Van's a used up man

These pictured presidents were object lessons, and
helped her to place many an incident in American history.

Another picture, one that pleased her more, was a
penciled sketch of 'Jeptha's Ride' (the work of that same
brother, Nehemiah). In it Jeptha was pictured strapped
to the back of a wild white horse. The fright of the horse
as he bounded through the forest over broken tree tops
was emphasized by the play of the powerful limbs, the
flowing mane and tail, and the dilated nostrils. It was
the horse, rather than Jeptha, that fascinated the little
girl. To her sorrow she does not know what became of
that picture, nor those of the Presidents, but she does
possess one other that hung in the spare bedroom, her
mother's work, a sampler of 'The Lord's Prayer' worked in red and green wool on fine perforated cardboard.
The lines are so spaced as to give room at the end for
ornamental loops of hair, cut from the heads of her
father and mother, sisters and brothers, ten in all; the
whole enclosed in a flowing vine of green and red. Beautiful work, done in some of those early, possibly homesick years.

There were other pictures that belonged to a period a
little later, but still within the life of the Little Girl.
Two small monochromatic paintings, one of moonlight
over an old castle on the river Rhine; the other the bold
crag of West Point jutting into the Hudson River. There
was also a fine steel engraving of Christ blessing the lit-

tle children, and there were two large oil paintings, one especially dear to the heart of Little Girl because it was of autumn tinted foliage with a shaft of sunlight touching the rocks and moss under a tall tree in the foreground. Many a time in the winter, she lay on her tummy and absorbed that pictured bit of color and sunshine. The promise is given that some time it shall come back to her.

The Curtis family, Lyman, Jennett and the two little girls, had been settled in the new house on the prairie since the twenty-third of April; the cabin in the woods was forlorn and empty, as dark and silent as the forest around it, but holding many memories of those happy years. They all missed the friendly shelter of the great trees, and the beauty that was underneath; but the prairie had a charm all its own, and left no room for regrets. Life in the frame house on the prairie moved on as contentedly as it had in the cabin in the woods.

Moving into the Prairie House

CHAPTER   SEVEN

The   Prairie   Years

Girlhood

Dawn the mornings clear and bright,
Blythely run the days,
Sunset colors tint at night.
Pleasant all the girlhood ways.
L.C.F.

WHEN the land for the prairie farm first came under
cultivation, a ten acre lot near the center front was reserved for the house, barns, stock and stack yards. A
substantial board fence set off the whole front of the prairie farm, and the rest of the one-hundred and sixty acres
was enclosed by a high stake and rider fence - a necessary protection in those days, for cattle roamed the prairies
at will. The board fence had two big gates. The entrance
road curved in at one side of the ten acre lot, passed the
house and barns, and went out at the other.

The flat tops of the big gate posts were ideal seats for
Mary Lizzie and Caro, and this was the place from which
they carried on their favorite summer game. After they
were perched up here with faces turned to the east, one
of the little girls would shout a loud "Hello"; and instantly from the woods came back an answering "Hello." The
game was on. Echo Man lived in the big woods and was a
fine playfellow, always ready with his answer. Any non-
sensical thought that popped into either of the two heads
was shouted out and tossed back to them. It was rare fun
while daylight lasted, and when daylight was gone, bed-
time was waiting. Two sleepy little girls almost literally

tumbled from the gate posts into bed. Nine o'clock was
bed-time in Father Lyman's house, a rule that was strictly enforced through all their childhood years.

The frame house on the prairie was built in the form
of an upright facing the east, with an ell on the south side
that was only a story and a half high. (The front door
opened into a hall from which the stairs led.) This front
hall had two doors, the one on the right opening into the
parlor, and the one on the left into the big living room of
the ell. The whole upright was unfinished save for the
two bed-rooms at the rear. These bed-rooms were entirely separated. One could only get from the parlor bedroom into the family bed-room by going the full length of
the unfinished parlor, across the hall, down the big living
room and into the kitchen, where a door on the north side
opened into the family bed-room. Neither were the upstairs rooms finished, save for a room at the rear which
was furnished for the hired man.

From the upper landing on the stairs, a low door opened directly into the half-story loft over the kitchen. It
was loosely but effectually floored, and had a wide window in its southern apex. Stove-pipes from both kitchen
and living room ran up to their chimneys through this
room, which was used for drying and storage purposes.
It offered an ideal playroom for the two little girls, when
winter would shut them in. The sunny warm loft was a
paradise for the rag dolls and the numerous families of
paper dolls that flourished here undisturbed, visited back
and forth, gave their dinner parties, - in short did everything that real people do.

Wonderful French costumes were evolved by the little
girls from the treasured bits of fancy paper, with a touch
of paint here and there and a bit of glue for joining. Mary
Lizzie never possessed a 'boughten' doll, but she loved
her big rag doll with all the strength of her little heart.
'Liza', named for the favorite Aunt, filled all the longing

and desires of a little prairie girl as much as any Parisian wax beauty would her city cousin. Each Christmas 'Liza' was treated to a fresh crop of hair - dark or brown or red, or even flaxen, saved from the summer corn, silks by Mother Jennett; her face and hands made all white and clean, her eyes sky blue, and lips and cheeks freshly pinked with paint. She was most certainly a beau-. tiful creation to gaze upon as she gracefully protruded from the top of Mary Lizzie's Christmas stocking.

In winter, when the mornings were bright and sunny, and their work was done, Mary Lizzie and Caro were ready for anything. They wore heavy shoes that laced high over long home-knit yarn stockings, red flannel knickerbockers lined throughout with soft cotton material and buttoned below the knee, red flannel underskirts, linsey woolsey dresses, hoods and long thick coats. All along the stake and rider fence of the home lot was a line of covering snowcaps, blown into shape and banked in hills and valleys. They were crusted into sparkling double-track roadways for the sole use of the two little girls who, with their stout shoes and flat boards, sailed on from the barn corner, up one bank and down the next clear to the front gate. Here they took to the other side of the fence and came racing home.

After the February thaw had passed, when Mary Lizzie and her sister were a few years older, the big slough was their playground. It ran for miles to the west, turned down just north of the Curtis farms, and made its way to the river, leaving a broad expanse of frozen pools with intersecting hummocks. The slough lay between the Curtis farms and the schoolhouse. Here was a chance for slipping, sliding, or skating. When the sun was hot enough, by four o'clock, it gave them fine exercise in the 'running broad' as they jumped from one little hill to the next over clear pools of icy water. Occasionally they sat down on the slippery surface or even in one of the icy

pools, but neither minded. Most of the water shook off
the flannel underclothes, and exercise did the rest.

The lean years were over all the land, but they settled
heaviest over the Iowa pioneers. Hardly was the moving
into the new home done, when Baby Kate came to gladden
the hearts of all. Her great brown eyes looked very so-
berly on life, and her lips trembled when anything went
wrong. She was born for love, and love she had in wor-
shipful quantity from everyone, but especially from two
little sisters. Even Father Lyman's stern rulings were
softened for her, though not in the matter of early rising.
He required all the grown ups to be up at five o'clock ,
and the children at five-thirty. However, Mother Jennett
insisted that her babies should not come under this rule
until they were five years old.

Catherine Smedley, she was named, with the pet name
of Katie. Mother Jennett called her 'her hard times baby',
since she came at the opening of those strenuous years.
There was now no money to buy cloth, so Jennett had to
clothe her out of the left-overs from fairer years.

Katie was a dear, and the whole family thought her
sweet and pretty. One of her charms was a tiny double
dimple - a "dinkle", she called it - which twinkled underneath her right eye. She had only to smile and the dimple
did the rest. She was early carried to 'meeting' just as
were all of Jennett's babies. But oh, this baby was not
the peaceful sleeping infant that Mary Lizzie had been.
As soon as the good Brothers in the 'Amen corner' lifted
up their voices in worship, Baby Kate joined them in
frightened wailing crescendo that no efforts could subdue
until Father Lyman had carried her out of doors and far
enough away for the distance to soften the singing. It was
the same thing over, every Sunday of her babyhood. Possibly her ears were so delicately attuned that the volume
of sound actually hurt as it struck them, for Kate developed later a rarely sweet voice.

Every Sunday night Mother Jennett gathered her family together for a song service that Mary Lizzie loved. Mother Jennett was herself a sweet singer, and took all the children under her training. Caro, Kate and George responded nobly, but poor Mary Lizzie, who loved songs best of all, could not sing a note. Though her voice was sweet enough, she could not be taught to follow tunes correctly. She undoubtedly was Father Lyman's own child, for he could not keep in tune, though he joined Mother Jennett and the children in singing the old hymns. One song gave him special pleasure. This was 'Old Uncle Ned', and when they sang it he would throw back his head and sing with great gusto:

> Dar was an old nigger
> And his name was Uncle Ned
> An' he lived long ago, long ago.
> He had no wool on de top of his head
> On de place where de wool ought to grow.

> Chorus
> Den hang up de fiddle and de bow
> Lay down de shobel and de hoe -
> Der's no mo' hard work for poor old Ned
> 'Case he's gone where de good niggers go.

> Old Ned had fingers
> Like de cane in de break
> And he had no eyes for to see
> He had no teeth for to eat de corn cake,
> So he had to let the corn cake be.

> Chorus

62

When Old Ned die, Missie
Took it bery hard, and the
Tears roll down like the rain.
Ol' Massa tu'n pale and he felt bery bad

Chorus

Many times the children of Mother Jennett have wished there had been recording machines for taking down
those old songs, their records are so hard to find in song
books. They were all lovely, from the old Scotch ballads
'Bonnie Doon' and 'Bonnie Dearie' down through 'Old Long
Island's Seagirt Shore', 'In the Michigan Forest', etc.

Bonnie Doon

Ye banks and braes o'Bonnie Doon
How can ye bloom sae fresh and fair
How can ye chant ye little birds
And I sae weary fu' o' care.

then

Bonnie Dearie

We'll go down by Clouden's side
Through the moonbeams spreading wide
O'er the waves we'll swiftly glide
To the moon so cheerily.

A PIONEER GIRL

Chorus
Ca' the youths to the kno's
Ca' them where the heather grows
Ca' them where the heather blows
My Bonnie Dearie

Ghost nor phantom shalt thou fear
My Bonnie Dearie
Thou'rt to love and heaven so dear
Naught of harm can come thee near
My Bonnie Dearie

Chorus

Again

Long Island's Seagirt Shore

On Old Long Island's seagirt shore
Many an hour I've whiled away
In listening to the breakers roar
That washed the shore at Rock-a-way.
To see young Iris as she dips
Her mantle in the sparkling dew
And chased by Sol away she skips
O'er the horizon's quivering blue.

Chorus
On Old Long Island's seagirt shore
Many an hour I've whiled away
In listening to the breakers roar
That washed the shore at Rock-a-way.

64

Transfixed I've stood while Nature's lyre
In one harmonious concert broke
And catching its Promethean fire
My inmost soul in rapture woke.

Chorus

To hear the whispering night winds sigh
While dreamy twilight lulls to sleep
And the pale moon reflects from high
Her image in the mighty deep.

Chorus

'In the Michigan Forest' had a fascination for us, perhaps because we must always shiver over the picture of
Indians and listen for Tecumseh's war-whoop.

In the Michigan Forest

In the Michigan Forest how dark was the night
When Indians and Britons were lurking for fight.
The trees, cliffs and valleys were pathless and cold,
And the dark waves of the Razine congealed as they
                    rolled.

In a moment so dreary what bosom could fear
That the Indians were lurking in ambush so near
And how could the night weary sentinel know
What bush held his brother or deadliest foe?

A PIONEER GIRL

Long, long shall Britannia remember that day
When the Indians were waiting in battle array
Till the whoop of the onset, the Chippewas raised
And lighted with canon, the wilderness blazed.

Then foremost in battle, Tecumseh was seen,
More fierce was his aspect, more hideous his form,
And louder his voice than the demons of storm,
"----- -------- --- --------------- ----"

My own babies cared most for the merry strains of
'Gypsy Daisy'.

Gypsy Daisy

Gypsy Daisy came a-tripping o'er the lea
She sang her song so sweetly
She sang till she made the green-wood ring
And charmed the heart of the lady.

Chorus
Raddle, raddle, dingo dingo dey
Raddle, raddle, dingo Daisy
Raddle, raddle, dingo dingo dey
She's gone with the Gypsy Daisy.

Lord came home late that night
Inquired for his lady
The servants gave him this reply -
She's gone with the Gypsy Daisy.

Chorus

How could she leave her house and home?
How could she leave her baby?
iiow could she leave her wedded lord?
To go with the Gypsy Daisy?

Chorus

Aunt Alice declared, when on one occasion in Florence's babyhood, she was left in charge for the evening, that only the strains of 'Gypsy Daisy' could subdue the wails of baby Florence. So she sang it over and over and on and on. She was ready to sleep herself - but not so the baby, so "Raddle, raddle, dingo, dingo dey" droned on until Florence's mother came to relieve the singer.

There was one other song specially loved by Father Lyman, and the Sunday evening 'sings' were never complete without it.

Mary to the Savior's tomb
Hasted at the early dawn
Spice, she brought, and sweet perfume
But the Lord she loved was gone.

For a while she lingering stood,
Filled with sorrow and surprise
Trembling while a crystal flood
Issued from her weeping eyes.

Ye who weep for Jesus sake
He shall drive your fears away.
What a change His word can make
Turning darkness into day.

There are many more strains from Mother Jennett's songs ringing down through memory's years, but only in snatches do they come and go, and most are lost forever.

Sunday night sings and corn popping

When Jennett wrote to her mother about Kate, she also mentioned that, "Caro and Mary Lizzie are good girls, going to school."

The first school, taught by Maggie Harwood, was held in the unfinished parlor of the Curtis's prairie home. The next was upstairs in Uncle Will's house, and it was during one of the sessions here that a big rattler was discovered lying along one of the rafters. The children

68

tumbled pell-mell downstairs, while Uncle Will was hastily summoned to kill the snake.

It was downstairs here that Mrs. Boynton taught Mary Lizzie's third term. She was a sleepy little girl, and would have tumbled off her stool had not her teacher lifted her off to finish her nap on her own bed. She had passed out from under Mother Jennett's rule that her babies should not be wakened so early as the grown-ups, and therefore she was up at five-thirty every morning. When the stern Father Lyman ordered,"Keep her awake," Mrs. Boynton met him cooly with, "Mr. Curtis, the child needs this sleep, and while she is under my charge she shall have it.", so Mary Lizzie slept in peace.

During these pioneer years, school was as much a part of life as anything else. There were two terms, a summer and a winter, each of four months. Those who had homes more nearly in the center of things vacated, each in turn, one of the precious rooms to make space for the teaching of their children, and teachers - the best obtainable - were put in charge. From Uncle Will's the school passed over to one term in the George Rozier house, and then on to the Strong homestead. It remained

School

here until the Union School House was built. This was on the County Line Road, about a half mile nearer to Mary Lizzie's home. It was called the Union School House because the ridge board was painted red, white and blue. Here Caro, Mary Lizzie and Katie went to school.

While the prairie years were still young, Grandfather Newman Curtis met his death, the result of a fall from a tree. He lived some three years after the accident, fighting hard for life and always hoping that he might live to see Grandmother and the boys more comfortably settled, but it was not to be. The plans for a new frame house were only partly completed when the end came, and he was buried from the old log house.

Grandmother Curtis was left with the three younger boys to run the farm. Lyman, the oldest son, was the natural executor of the estate, so that heavy care was laid upon his shoulders. He was also guardian for the two younger boys, Henry and Wilbur. But the prairie farm in Bremer County was still home for Mary Lizzie and her family, and would continue to be so as long as Grandmother and her boys could manage the farm at Independence. Here the new frame house was built, good board fences put up; and to Mary Lizzie's sorrow a beautiful grove of wild crabapple trees just to the rear of the house was grubbed out to make room for the new farm buildings.

Long and low and gray was the house on the Curtis farm, but it held a merry crew of young people. It was from this same frame house that Mary Elizabeth and Elinor were married. Long and low and dull and gray that house may have been on the outside, but never for one moment could it have been dull on the inside, for many young fellows came courting the pretty Curtis girls, both true to the Curtis type - blue eyed and fair.

70

Of the weddings Mary Lizzie has no memory, but Elinor married Mr. Leonard Hart, an Independence lawyer. "Uncle Hart," we always called him, and Aunt Elinor always addressed him and spoke of him as "Mr. Hart." They lived the better part of a life time in Independence and raised their family there - Frank, Fred, Monty and Lillian, but in the end Uncle Hart and Aunt Elinor both died in Texas. Their first born, little Myrta, loved best, first and always by Mary Lizzie, died in Independence.

Mary Elizabeth, the youngest of the Curtis girls, married a Mr. John Whaite. She, too, called her husband "Mr. Whaite." Mary Lizzie was very fond of this Aunt Mary, and grieved over her early death.

Three boys, Lora, Henry and Wilbur, were yet a part of the home circle, and so also was Newman, the handsome young bachelor who loved all the girls so well he could never decide which one to marry. This is his own explanation for the state of single blessedness in which he lived and died.

Mary Lizzie liked the long low house that held her merry young aunts and uncles, and Grandmother Curtis with her soft white lace cap, her cane, the caressing flutter of her hands, and her tender greeting - always the same: "Pretty Cretūr, Pretty Cretūr."

Grandmother Curtis had not always been the gentle, drooping figure that Mary Lizzie knew, for she bore and raised fourteen children. Those eight boys all grew into strong men, men of uprightness and ability, and six girls kept them company. Uncle Newman once told Mary Lizzie a tale of Grandmother, patience gone, chasing the young rascal with her broom round and round the outside cellar until she stopped, exhausted; when he whirled, caught her up, and carried her on his shoulder into the house. It all ended in a gale of laughter, but the boy meekly obeyed his little Mother's command, "Newman, you do as I told you." They all respected her wishes, from Lyman down.

It was with this same dear eldest son, Lyman, that she spent her last days, rounding into peaceful completeness that full life of more than ninety years. When the Angel of Death stooped over her, she sat at Lyman's right hand, eating dinner and chatting with the family. So gently the summons came that her spirit fluttered out between Lyman's question and her answer, leaving the smile of her unspoken reply to remain as a tender memory in the hearts of all around her. The Angel that bore her spirit upward knew the loving greeting that awaited her, "Enter thou in."

Even though times were hard, settlers came from New England or the eastern states into this special locality. Lyman's brother, William, owned the next farm to the north, while a mile and a half to the south, brother Lora had his land. But the great stretch of prairie was still untouched by the plough, for men generally sought the shelter of the timber.

In August of that year Lyman wrote to Jennett's brother, Nehemiah, in regard to the hard times: "Brother, what is the prospect for you all in that northern region? I hope you have some crops to harvest this summer. As to ourselves in Iowa, the crop of small grain is now known to be an entire failure; even the oat crop will not pay for harvesting and threshing. Hence the future looks gloomy and bad as the times are now. And I assure you I never knew anything to compare with it in our region. Money is almost out of the question; yet I think the year to come will be more distressing, for I fear a want of bread. Wheat with us, which four weeks ago could not be sold for twenty-five cents cash, is now held firmly at one dollar per bushel, and some seem to think it will reach two dollars before spring." (Which it did.) "Yesterday I saw and conversed with a gentleman who had just returned from a trip through Illinois, Indiana and Ohio. He assures me that the crop everywhere is almost an entire failure, except corn, and that - even under the most favorable circumstances - cannot exceed half a crop. The potato crop with us - even at this early day in the season - is exhibiting evident signs of immediate decay.

"Now with this sad prospect in view, perhaps you would like to know how it is likely to fare with your brother and sister and their little ones. And I am happy to tell you that I have a good barn with grain in it sufficient for ourselves and our beasts for another year, and some to spare for those who shall be in want, for which we hope to be truly thankful to the Giver of all Good. I think this is the case with about one in ten through this region. Some persons have recently come in and are going back to Illinois for the winter. I fear they will not find it much better there, but it will help those who are left behind."

It was a time of fluctuation in prices. During one of the pre-war years, Lyman's oat crop on the prairie farm was cut and stacked, when prices went so low that he did not think the oats would pay the threshing bill, and so he let them stand through the winter. His work was so well done that he knew his stacks would weather the storms and the oats would come out all right in the spring. When the next spring seeding time came, they were selling at the fabulous price of three dollars per bushel, and brought him a round thousand dollars. He always considered this his real start in farming, for it gave him the needed ready money to carry on.

Life was strenuous, but Mother Jennett was of heroic mold, and three times during those prairie years she went down into the depths to bring up little lives. Next after Katie came the brother, George Washington, for whom Mary Lizzie and Caro had sighed; and still later

73

A PIONEER GIRL

came the twins - two waxen baby dolls, a boy and a girl,
who fought for the breath of life but had not the strength
to carry on. So they were laid by the mound where little
Willie slept, and life moved on without them.

CHAPTER EIGHT

The Prairie Years ... FIRE!

The Prairie Years

I climbed the high mountains,
Sailed over the sea,
But the song of the Prairie
Went ever with me.

L.C.F.

MARY LIZZIE loved the spring on the prairie. Grass was growing, flowers were pushing up, meadow larks singing and robins piping. Quail called across the spaces and night hawks circled and zoomed. In the fall, golden rod and asters stayed until the sumac began taking on rich coloring and hazel nuts were ready to open. The grain fields that were perfect levels of green in the spring were now gold and russet. There were nests in the grass which had held speckled eggs in the springtime but now were empty and deserted.

The corn stood up straight and tall, "Laid by," Lyman said of that big cornfield, and there was satisfaction in his tone, for no more cultivation would be required to grow the heavy ears and ripen them off. On a hot summer night, one could listen closely and hear the rustle and crack of fast growing corn, as the kernels swelled and the ears grew big and heavy. Harvest time whirled on, and now it was Mary Lizzie's task to carry lunch to the harvesters and to keep the water jug filled.

When she went for the cows, Mother Quail, with her dainty crested head held high, stepped leisurely before her, while her little flock scattered from under her feet.

75

Sometimes Mary Lizzie was quick enough to put her hand
over one of the baby quail and cuddle it for a breathing
space against her cheek, before she put it down to scurry
after the others. Sometimes a rattler coiled near where
she passed, and sounded his warning whirr to send her
flying on.

One fall, for an unforgetable space, an army of thousand legged worms in their migrating crossed the prairie.
They filled the County Line Road from side to side. For
weeks the wheel ruts were packed with a horrible wrig-
gling mass of reddish brown, many-legged, wholly nasty
worms. Mary Lizzie's bare feet would balance on one
hummock, spring to the next, and so on to the cows and
home again. If she missed her hummock, which she did
many times, Ugh! Fortunately the worms had not come
to stay - just crossed the prairie; but that was bad enough
and lasted quite long enough.

Mary Lizzie easily forgot the horrible, and swung
along happily through the gorgeous fall days. There was
stacking of the grain, and many a ride on top of the load-
ed rack as it journeyed from the field to the stack. This
always ended with a thrilling leap from the top of the load
to the floor of the stack, and from there Mary Lizzie was
swung out by her father's strong arms to the ground.

Father Lyman's stacks were a joy to the eye. There
were no lop-sided or squat deformities in his stack-yard.
The bundles were laid with precision in a perfect circle,
all butts to the outside and heads to the center, row upon
row, in graceful, steady slant from the bottom bulge to
the single top bundle. The long ricks of barley, of tim-
othy, or of wild hay were formed just as symmetrically.
As soon as the stacking was done, the big horses were
put before the plow running first a band of safety furrows
around the stack-yard and house. Next, long furrows
were laid on the prairie side against the western fence --
enough to serve as a guard from which to back-fire in

76

case the dreaded holocaust should sweep down.

Once when Mary Lizzie was still small, a prairie fire did come rushing out of the northwest fast and furiously. The crisp fall morning held a taint of smoke, enough to make Father Lyman scan the north anxiously; but he saw no sign, and there was no wind, so the big horses were put before the wagon and he hastened off to his job of threshing at a neighbor's some three miles to the south. But as a precaution the cattle were left in the barnyard. Jennett could see to feed and water for them.

Slowly the smell of smoke grew stronger, a brisk northwest wind blew over the heavy slough grass, and the air became unbreatheable, the cattle lowed and galloped around the barnyard, and chickens sought their roosts. The horses snorted and stamped in their stalls. All Nature was afraid. Windows and doors were closed to keep clear space under the heavy bank of smoke that was settling over the prairie landscape.

Mother Jennett watched anxiously for Lyman, but while she waited, with Caro and Mary Lizzie to help, she worked steadily at the windlass. The drinking trough and the half barrels on the pump platform were filled, the empty rain water barrels were whirled into place; tubs, the kitchen boiler, kettles, pails and pans were filled. Broom, mops and every available piece of heavy old cloth were used to fight the fire. Not much was left of Jennett's carpet rags. Matches were at hand. and a lunch fixed that could be eaten on the run. Already a faint but ominous crackling sounded, and the air was filled with an all pervading roar.

Three miles further away from the onrush of the fire, the workers at the threshing machine, as the smoke pall reached them, separated, each to protect his own; but first the united teams pulled Lyman's machine into the center of a ploughed field where it would be safe. Then the whip-lash curled over the backs of the big horses with

Lyman standing on the buckboard, and snorting and plunging, they took the road home at a dead run. So heavy was the smoke pall that the watchers could not see them until they rounded the corner of the ice house by the well.

Lyman threw the lines to Jennett and said, "Tie the horses in their stalls, but leave them harnessed; we may have to use them later." So they were tied in their stalls where they might eat but be ready for instant use. Lyman seized his two pails of water, matches and his fire-fighting cloths, and rushed off to backfire along the west fence, inside of which was ripened stubble leading to house and barns.

The horses attended to, Mother Jennett and Caro, each carrying fire-fighting paraphernalia, were off to join Lyman at the fence. Mary Lizzie could be trusted to take care of her small sister Katie and her baby brother. She never forgot a single incident of that terrible night.

The air was full of charred fragments of grass, carried by the wind and dropped for Mary Lizzie to put out. The crackle and roar were deafening. Prairie wolves scurried into the enclosure; swarms of birds, lost and frightened, circled through the air, calling pitifully; prairie chickens scuttled into the security of the lot; and quail flew down in swarms to rest and breathe.

Off there where her father, mother and sister were battling was a solid wall of fire. The flames swept over the heavy dry grass of the big slough, each especially tall, ripe bunch sending them leaping and crackling skyward, until it seemed to Mary Lizzie that the whole world was on fire and everyone burned up. Every now and then one of the fighters would run in for more water, a drink or a bite of something; then snatch up the pails and go back as fast as breath would let him. While daylight lasted Mary Lizzie's small arms toiled patiently at the windlass, to raise and empty the heavy bucket, striving with all her might to keep up the supply of water. There was one blessing. The wind lifted and carried the smoke pall on, while the flames danced and crackled and threw their wierd red light over everything. It was an inferno such as Dante alone could picture, and for Mary Lizzie the night stretched into infinity. Mother Jennett came back to nurse her baby and put him to bed. Kate slept where she lay covered on the floor.

For a time Mary Lizzie flew from the sleepers to window or door and back again. Then, as her fears increased lest when the flames came - and she knew they would come - she forget that baby brother, she drew mother's rocker close to the sleeping Katie and, gathering Georgie into her small arms, sat and held him through that endless night. Oh, how many hours there were in that night; how long it was, and how her back and arms ached! She dozed, to waken with a frightened clutch at the sleeping baby. Would mother, father and sister never come?

A PIONEER GIRL

Cassabianca at his post on the burning ship had nothing on Mary Lizzie. An exhausted sister Caro was sent in first, so dead tired she could hardly pull off her clothes before she tumbled into bed. Then Mother Jennett came, and all was well with Mary Lizzie. She, too, was sent to bed. A sleepy Katie was undressed and put beside her, and she knew nothing more until the sun was shining.

Father Lyman had not only his own to protect, but also Uncle Will's farm that joined on the north. Here had been the hardest fight, for it was now rented property, and rows of tall weeds led from the boundary fence on the west down by the pasture fence to house and stack yard. The fire must be stopped on the outside, lest all of the two farms go. So he had fought and watched in case some puff of wind should fan charred embers into flames, and fire again rush forth. His watchful care saved both farms and fences. Not all the prairie dwellers were so fortunate, for the furious gale sent the flames leaping over hastily thrown barriers into the tinder of stubble fields that led to stacks and homes. Only those who had looked ahead and planned protection long before the fire came, saved their homes and stack yards.

That night went down into prairie history as the night of the great fire, and the morning showed black desolation. Thankfully the Curtises reported, "No lives lost, some stock and much property, but all our people are safe. God be thanked."

It took years to build back and replace the heavy toll the fire had taken. Some never did, and the care of the less fortunate neighbors taxed still heavier the powers of the more provident ones. Yet they all came through these lean years by hard work and strict economy, and in the end conquered and were surrounded by peace and plenty.

Many neighbors came to Lyman for help, since he was known throughout the prairie country as a successful farmer, a man of means and power. Time and again he was urged to run for the State Legislature, but he steadily refused. He had no political leanings.

Before the prosperous years came, there was one never-to-be-forgotten Christmas when Caro and Mary Lizzie, little as they knew of the desolation abroad over the land, yet feared that Santa Claus would not come. On this Christmas Eve Mother Jennett met two sober faced little girls with the cheery assurance, "Of course he will come. Old Santa will never fail you."

"Will Santa Come?"

So they wished their Merry Christmases, hung up their stockings, and went happily to bed. Sure enough, when Christmas morning came the Old Fellow had been there, and the long stockings bulged with nuts, doughnut boys and girls, long twists of home-made candy and pop-corn balls - whose making had kept Jennett up into the 'wee sma' hours.' Those pop-corn balls were a rare delicacy.

Only the tenderest and most fully flaked kernels were used, and over them went the thread of hot molasses candy, cooked to just the right stage. Then the hazel nuts were dropped in, handfuls of them, until each fluffy ball, worked into shape in Jennett's hands, was full of meaty goodness.

The rag dolls were fresh and clean from the inside out, and stood gloriously forth in new outfits.

There followed the Christmas dinner with Uncle Will's family, and the play with the three cousins, Herman, Hattie and Annie.

It was in the prairie years that the terrible winter still spoken of as the worst that Iowa ever knew came and went. So intense was the cold on that New Year's night that stock froze where they huddled, and ears and noses were frosted while people slept. In the more provident homes, fires were kept roaring. Katie was tucked in between Caro and Mary Lizzie, who were sleeping in the far parlor bedroom, well covered up. As the cold deepened in intensity, it snapped and cracked in the framework of the house, and Mother Jennett made many trips to the far bedroom to make sure all was well with the sleepers. Finally she pulled a feather tick from an unused bed and laid it carefully on top of all. They wakened in the morning with ears, noses, fingers and toes all safe from the frost.

Lyman saved his live-stock by bundling up and going out among them. With heavy work whip he routed every head up, and whipped and chased them round and round the barnyard until their blood was circulating vigorously. Then he took his basket of corn and ran before them, feeding each one ear by ear until all were chewing contentedly, before he left them to settle down in deep beds of straw. The horses in their stalls had an extra supply of bedding and were covered with blankets and given oats

to keep them munching. There was little sleep for Lyman
and Jennett through that terrible night, but all were safe
and sound when morning came.

The cold night when Lyman kept his cattle on the move

CHAPTER   NINE

The Prairie Years Continued

At Meeting

THE Chitisters were the nearest neighbors. West and
north of them were the Strongs, the Boyntons and George
Rozier, an eccentric sort of hermit, thought by some to
be a little crazy. Across the river and some miles farther down was the Sterling home. Instead of the black
soil of the prairie, here was a sandy loam that grew wonderful melons. In melon time there was always a visit at
the Sterling home; a wonderful playtime, and all the
melon one could eat. The return visit of the Sterlings
was usually made in the winter, when Libbie and Caro
and Mary Lizzie played with rag dolls and paper dolls in
the warm loft over the living room.

Straight across the river from Lyman's about three
miles east, was the Older home, and near it but still
farther away clustered the homes of other settlers. A

schoolhouse, known as the Older School, had been built here and in it was held Church and Sunday School.

Sundays were red letter days for these pioneer children. Not only did they wear the precious Sunday clothes (no year was so hard but that each child had her Sunday outfit fresh and clean, and regularly as the day came around she put it on), but in addition there was prospect of subdued though none the less happy visiting among the junior members of the families that gathered here for worship. Libbie Sterling, Emma and Clarence Older, Joe Chitister, John and Ellen Strong, Addie Boynton and a possible half dozen others were the mates Caro and Mary Lizzie met here.

Rarely did the family miss Church and Sunday School. Mary Lizzie went to Church during the Forest Years carried in her mother's arms, but as soon as the long dresses were put away she was deposited with Caro on the soft bed of clean straw in the bottom of the big lumber wagon. In those first years she slept most of the trip.

As she grew out of babyhood those drives to and from the Older Schoolhouse, where 'Meetings' and Sunday School were held, were to her a never failing source of delight; a mile and a half across the prairie to the woods, then under the trees for another mile and a half to Chitister's Mill, where a bridge crossed the Little Wapsi, then on for five more miles, past rolling hillsides interspersed with groves and bordered with ferns and flowers.

Once they saw a man stretched on a bed of moss near the roadside, hands locked under his head. He lay perfectly still, only turning his head to watch them pass. It was a sight strange enough in those days to fasten itself in Mary Lizzie's memory for life. She still wonders who that man was and what he was doing there.

Church, and Mother Jennett's after Church Class meeting were as great a trial to Mary Lizzie as the ride was a pleasure, but it had to be endured for Caro and Mary

Lizzie were required to sit quietly through the two sessions that followed after the Sunday School. Sunday was
the Lord's Day, and no levity, or even seeming levity,
could be indulged in, and the children were required to be
most decorous. Once when Mary Lizzie sat down for a
quiet swing, Grandmother Curtis's compelling, "Lyman,
that child ought not to swing on the Lord's Day" brought a
command, "Get out of the swing, my child," and Mary
Lizzie got out - and quickly too, for she had been trained
to obey without question. It therefore is not strange that
Sunday seemed a day of interminable length.

The commandment, "Thou shalt not work,..." was
obeyed literally by the pioneers; there was rest for both
man and beast. On Saturday night the farm machinery
was put under cover and the oxen and work teams turned
loose. The good people who gathered for services gave
the Lord his full share, and then gave themselves up to
the pleasure of the weekly visits between neighbors and
friends. Often there were Sunday visits at the different
homes, and while the elders rested and chatted with becoming Sunday gravity, the children were turned loose to
play. Anything that came to hand served as a plaything,
and every little variation in the everyday course of events
was a fresh delight.

A Sunday visit in the home of the Sterlings and the return visit in Mary Lizzie's home were pleasures to which
she eagerly looked forward. The one child in the Sterling
family was a little girl of her own age, Libbie, and happy
minutes filled the whole day. They wasted no time on
dolls - there was too much to be seen, and too many
places to be explored. When they were tired, Libbie
would get out her bundle of 'quilt pieces', and the children would go over them with delight, carefully smoothing
out each piece for its place in the bundle or bundles.
There was always a generous division, and Mary Lizzie
would go home at night the happy possessor of many new

pieces for her quilt. When the return visit of the Sterlings came and Mary Lizzie was hostess, there was the same generous division of treasured 'pieces.'

Quilt making was an art practiced by the girls' mothers, who used a variety of patterns - some plain and easy like four squares, and some really beautiful and intricate. There was one made up of hexagonal blocks used for wool or silk. A pattern was first cut from stiff paper, usually supplied by old letters, pinned to the goods. The cutting left sufficient margin to fold over the paper where the edge was carefully basted down. Those different blocks were then overhanded together around a central one. After all the blocks were put in place, before the quilt went over the lining, the basting threads were cut and the papers taken out.

The quilt that Caro and Mary Lizzie made was of wool pieces, but Mother Jennett had one of silk, the pieces no larger each than a silver dollar. These old quilts are treasures in memory's house, for so long as one block survives, she who pieced it can tell whose dress it came from, and before her mental vision flits a picture of the one who wore it.

Mary Lizzie and Caro were taught to sew on an easy pattern. They learned to thread a needle, to wear a thimble, and - hardest of all - to take the tiny straight stitches required of them. It was especially hard for Mary Lizzie because she was left handed, and Mother Jennett would insist that the needle must go between the fingers of the right hand. This she finally gave up, and consoled herself with the reflection that left handed quilters were always in demand; fitting into spaces where right arms would interfere with the next quilter.

The amenities were never forgotten in this household. Plain and severe as were the surroundings, Mother Jennett and Father Lyman held these little girls as strictly to their manners as they themselves had been held in those far away eastern homes. The family was always gathered around the table before Father Lyman began his serving. Frowsy heads were not allowed there, and the children were expected to take their seats when called and sit without slouching on table or chair. It was a stern and never ending drill.

There was work for everyone, even the children, in the struggle for existence in those first years. Mary Lizzie had her allotted tasks. The wood-boxes had to be filled, the pan of chips put under the kitchen stove to be dry and ready for quick kindling, and the cat fed. She began washing dishes while she was still so small that it was necessary for her to stand on a stool in order that her short arms might reach the dishes in the pan. Sister Caro did not like dish washing, and Mary Lizzie was easily coaxed into doing it for her when Caro's turn came. Caro had beautiful hands and she did not like to put them into greasy water.

There was one thing Mary Lizzie had to do that she always dreaded, for she was afraid. Her daily duty during the season of heavy work on the farm was to hunt up the cows, round them out of the common grazing herd and get them into the barn yard in time for milking. Mother Jennett did all she could for her by locating the herd and starting her off right. Then, armed with the buggy whip, out the scared Mary Lizzie would go, all by herself. She had no dog to keep her company, but the slender hickory whip stock with its long braided leather lash did much to bolster up her courage.

She loved the prairie and the tall gum weeds with their nodding yellow heads, where after the cattle had nipped them, came the drops of yellow resin that furnished her winter supply of gum - the finest ever. She loved the great splashes of pink and lavender of the wild prairie pinks and the dashes of vivid color of the tall lilies. Every knoll was crowned with pink roses in spring time, and red with their gay bolls in the fall.

Much as she loved the flowers, however, Mary Lizzie scanned these little rises in the level prairie ground with fear and trembling, for occasionally a lone wolf would wait here to watch her hungrily when she passed, circling wide with a great snapping of whip and with loud yells of defiance. Too much of a coward to attack that whirlwind of noise and whip lash, he would follow at a respectful distance until the little girl was safe among the sheltering horns of the cattle. How fast her heart beat and how relieved she felt when she ran among the cows, made no difference. her work must be done and it was done, and done on time.

Mary Lizzie drives the cows

Principles of duty and obedience to law were being instilled into her by the insistent demands of her New England father. Mother Jennett's heart was softer, and she would have saved her from the harsh rulings many times if she could have done so honorably. Honor was the key note of life to both parents, and the children could not fail to absorb it.

Brother George was the idol of the whole family, but especially of Mary Lizzie. As soon as his short legs could travel he was riding 'horse' astride some stick, and while he was still a crawler he would cry for his father's trousers. It was Mary Lizzie's pleasure to put the pair of trousers on a chair and lift his chubby body into the seat so she could crinkle those long legs up over the short ones to the delight of them both. But it was rough on Father Lyman's one good pair of trousers, and Mother Jennett had of necessity to stop it. When the walking stage was reached it was the same thing over, only it was in Father Lyman's Sunday boots that he stumped around the house, breaking down the tops of fine leather until Mother Jennett had again to forbid this and save the boots. The small, active brother must have something to keep him busy. The highly prized Henry Clay cane next attracted his attention, and like all small boys, what he wanted, he appropriated.

He was mounted on this cane when he disappeared one summer day. Whether Mary Lizzie had become so absorbed in her paper dolls that she failed to miss him, or whether he had wakened from a nap and gone off on his 'horse' unobserved by anybody, she did not know, but he was gone.

Wildly the whole family searched, - stables, barnyard, the house lot as far down as Uncle Will's, but he was not to be found. Then someone's distracted eyes - Mary Lizzie's, searching the far off places, saw a speck of brightness that might be his red waist, to the west just on the border of the big slough. "There he is." Arguing with herself as she ran, "He could never have gone so far," Mary Lizzie went flying toward that bright speck, hurrying, hurrying, lest it be swallowed up in the tall slough grass. Over the high rail fence, on and on she ran breathlessly, wildly calling, "Georgie," for now she was near enough to see that it was he. All the horrors of possible pools of water in the slough and of prairie wolves and rattlers lent wings to her feet.

Overtaken, a tired baby cuddled in her arms and said he was "des doin' to see Aunt Liza."

CHAPTER TEN

Still The Prairie

The War

Great God of Battle, lead them on
Those dear loved sons we gave –
They cannot fail who righteous march –
World liberty to save.
                    From "A Paean of the World War"
                              by L.C.F.

HENRY, one of Lyman's younger brothers, was a part
of the family on the prairie farm for a year or two, but
he soon entered Upper Iowa University, where later Wilbur followed him. Steadily the war cloud thickened, and
nearer and nearer came the inevitable outburst. Henry
was in his senior year when those first fateful guns were
fired on Fort Moultrie. Their blaze flamed over the
whole north; the country sprang to arms. A company
was enlisted in the Upper Iowa University.

No more of college for Henry. Books were cast aside.
He was still a minor and could not be accepted without
Lyman's consent, so he walked the eighteen miles to the
prairie farm to get that consent. The two tall men made
a never-to-be-forgotten picture as they stood before Jennett while Henry made his request. Lyman's hand was on
his shoulder and tears were in his eyes when he answered, "Go, my boy, and God bless you," to which Jennett
bowed a tearful assent; then she hurried her dinner while
Henry rested, and they all talked at once.

After dinner Jennett hastily covered the small pillow
from the cradle with a piece of brown broadcloth and gave
it to the tall boy who was going away to war. It came

home on furlough when Henry came, and re-enlisted when he did. Jennett re-covered the little pillow, and it went back to finish the war with that 12th Iowa Infantry. It survived all the marching, all the fighting, all the months of prison life, only to be stolen from under Henry's head as he slept on the deck of a river steamer after he had been mustered out. Those five years of faithful service deserved a more honorable ending - but the time of need was past.

When dinner was over, Lyman put the buggy team - Betsy and Billy - before the democrat and took Henry back to Fayette to send him to war.

They say goodbye to Henry

They found the university in turmoil. The older students had enlisted and many of the young professors - leaving the university almost without students in its upper classes. They had donned the army blue, shouldered guns and marched away, promising to return for graduation soon. Few if any of them ever saw the inside of the U.I.U. again. Some were left on the field of battle, some returned - but no longer young, no longer with the zest for college. Those five terrible years had drained out

93

the vigor and enthusiasm of youth. College years were gone.

Henry marched in Company C of the 12th Iowa Infantry through the full five years of the war. For six months he was a prisoner at Atlanta and in Richmond's dreaded Libby Prison. When Union prisoners were exchanged, they were sent home on furlough to rest and recuperate. A thin, white, half-starved boy was Henry, and the small-pox had done its worst. It seems almost a miracle that he escaped from that prison and that disease alive. But how he was feted. Friends could not do enough for him. No wonder he went back to his regiment restored to health and strength. And at the close of the war he was muster-ed out, alive and whole, 'Honorably Discharged.'

Wilbur had also enlisted as soon as they would take him, in the 3rd Iowa Artillery, where he served until the end. He returned to us a pale, weak stripling, but whole and still able to build up to healthy years of manhood. I wonder if it was possible for those at home to be thankful enough that the war was over, and that we had our own again.

To show the bitter animus of the times, Henry told that when marching through the streets of Atlanta to the prison stockade, women who looked as though they might be highbred ladies came out to spit upon them - these northern prisoners. Perhaps they had greater cause for bitterness than we, for not only was the fair flower of southern manhood lost, but their homes were gone, their land a wreck, and hope laid low. There was reason for bitterness, even though theirs was the fault.

But the business of life must begin. Henry had decid-ed to make law his profession, and he went into the office of his brother-in-law, Leonard Hart, to study.

When Uncle Neil, Jennett's brother, enlisted, Lyman wrote to him at his home in Dubuque: "Send Anna and the two children to me. I cannot get away to go, but I can and will take care of your family while you are gone." So the Dubuque home was closed and Aunt Anna, Nettie and Lettie came to the prairie home. Merry were the times that followed for the crew of youngsters at the farm.

Lettie and Nettie came to the farm

The four little girls went to school in the new Union School House. When winter snows were banked high, it was too far to walk, at least for the short legs of the two little girls, Lettie and Katie, so the children were taken to and from School. They were told to stay in the School House until someone came for them.

One day when Father Lyman was away, Mr. Boynton was to bring the children home. He lived in Uncle Will's house, and his own little daughter, about Katie's age, was with the children. For some reason he was very late in starting, so late that darkness was setting over the School House where the children waited by themselves. At last the patience of the two older girls gave out, and they started with the small sisters to walk home through the snow banks. In fact they walked or waded the whole distance and met the sleigh just leaving the Boynton home,

95

when darkness had fairly settled down. Father Lyman was just back from his trip when the children came in. There were two frightened mothers because of the lateness, and they were very angry and indignant at Mr. Boynton. They tried their best to save Mary Lizzie, who had disobeyed, but to no avail, and she had to take her punishment. She got comfort from the fact that Mr. Boynton was hauled over the coals for his carelessness.

The winter was a pleasant one, and the children were especially excited by the entertainment that closed the school, a little play, "The Red, White and Blue," with Nettie and Mary Lizzie draped in flags and red, white and blue bunting, impersonating America and the Goddess of Liberty.

In the years of eighteen hundred and sixty one to eighteen hundred and sixty five, Lyman and Jennett were working hard to bring the prairie farm under cultivation. Help was not to be had. The men had all gone to war. The fields must be tilled, sowed and harvested. The men in the army had to be fed. Only women and children were left to do the work, and they carried on bravely. Jennett had to do the necessary baking and cooking, but when that was done she took her place by Lyman in the field. Caro was old enough to work with her father and mother, but baby George and Katie must be cared for, dishes washed, beds made, and vegetables prepared. This was Mary Lizzie's work.

Lyman cut his hay and harvested all his grain with scythe and cradle. When the seasons of haying and harvesting were on, he worked early and late, swinging his heavy implements as he walked back and forth, back and forth across his fields, cutting a perfectly straight swath. When it came to the reaping, the old Kirby reaper was on deck, but Lyman's scythe first trimmed all the edges and laid the field off into regular headlands, so that no grain should be trampled by the horses feet. The swath cut

around the field was called a headland. Caro took the
driver's seat and handled the lines so Lyman could give
his mind to the sickle and to the spacing of the dropped
grain for the bundles. Only enough grain was cut each
day for him to bind and set up, and here Mary Lizzie's
turn came. She went before him and put the bound up
bundles into place for shocking, six and six, butts to butts,
in two rows with a path between.

Lyman walked through this center path from one set of
bundles to the next, lifting the side ones into place as he
went, and twisting the end ones so securely into a covering for the shock that his grain was never wet.

Once poor Lettie, in running over the field of stubble,
met the fangs of a rattler just above her shoe top. So
viciously had the snake struck that he had to be pulled
loose. Lyman carried her into the house; the shoe and
stocking were removed and the swollen leg steadily bath-
ed with strong soda water, while the little girl had cup after cup of buttermilk almost poured into her. The treat-
ment was a perfect success, and it was not many days before she was running around as gaily as usual. Probably
the stocking warded off part of the virus; at any rate the
snake bite, considered a deadly poison, left no bad effects,
though the spots where the fangs struck showed for many
days.

Busy and tired as they were, Lyman and Jennett found
time to follow with deepest interest the current of the
times. The Albany Weekly Evening Journal, with Thur-
low Weed at its head, came regularly into the home and
kept them in touch with their old world and the vital questions of their day.

Caro and Mary Lizzie had a special magazine,'Merry's
Museum,' and the most important part of it for them was
a few pages devoted wholly to the soldiers - letters to
and from and concerning them. The war spirit was strong
in children as well as grownups, and Mary Lizzie and

Caro had their full share. They would dance about, holding aloft the tallest and straightest cane that their short arms could manage, yelling at the top of their voices, "Hurrah for Abraham Lincoln," and then they would bestride a deformed stick and switch and generally belabour poor old Jefferson Davis.

Mother Jennett's busy hands found time to make a fine dress shirt. The clusters of small tucks down the front, stitched in by hand, followed accurately the thread of linen, and her buttonholes were perfect. When complete and beautifully laundered, a free will offering to the soldiers, it was sent to one of the Ladies' Aid Fairs in Dubuque, where it sold for three dollars. Later she pieced a quilt from scraps of silk and ribbon, cut in small hexagonal blocks about the size of a dollar, and when finished in every detail it was sent to the "Sanitary Fair" at Dubuque, another offering for the sick and wounded soldiers.

The quilt was a beautiful one, and of great value to the family on the Iowa prairie was the association, for each block brought up the memory of a happy face or a jolly time. It tugged at the heart strings so strongly that Lyman wrote in to the ladies and gave them the price they had placed on the quilt and had it returned to the home farm. I think that price was fifteen dollars.

These pioneer women devoted one afternoon of every week to a neighborhood gathering where they worked for the soldiers. Socks were knit, bandages were made and rolled, and every scrap not otherwise used was scraped into the lint so necessary for the wounded. This scraping furnished work for little girls like Mary Lizzie. They also made and filled the pincushions for soldiers' comforts, as the compact little bags were called that held courtplaster, scissors, thread, pins, needles and buttons. There was work in plenty for the girls, and they had merry times in doing it.

Henry and Wilbur had gone to war. Grandmother Curtis was old and feeble and no longer able to manage the farm. Lora and William both had families and farms of their own. Each in turn tried for a year or so to run the estate, but so unsuccessfully that each dropped the job and went back to his own farm. Lyman was guardian for the younger boys and had to look out for their interests as well as for his mother's comfort, but this was so hard for a man who lived twenty-five miles 'up the river' that it became necessary for him to rent the Bremer County farm and take over the management of the estate. The prairie years were ended.

In January of eighteen hundred and sixty-five, as Lyman and his family drove into the yard of the new home, they heard the Victory cannon booming. One by one the cannon told off the victories of the last days of the Civil War: Richmond had fallen, Lee had surrendered, Jefferson Davis had been captured, and at last came the tumultous booming that told the end had come. The war was over!

Mary Lizzie was twelve, and responded joyfully to every boom with the throb of her whole being. If she could not appreciate what peace meant, she could and did exult to the fullest in that glorious victory. Victory meant to her the defeat of the rebels and the final flaying of 'Old Jeff Davis.' Even a child of her years imbibed the spirit of the times.

Triumph at end of War

A PIONEER GIRL

Although the fighting was over, the preliminaries were
not settled, and peace was not actually declared until
March or April. It was even later in the spring before
the boys came back.

Then came that terrible day when bells tolled and flags
hung at half mast. Lincoln was dead! Struck down by an
assassin! The whole world shuddered and was stilled;
cheeks paled and eyes grew tragic, as with bated breath
people listened to the ominous toll of the bells; with bowed heads and anxious hearts they stood in silent sorrow.

To add to the solemnity, while the bells tolled, an
eclipse was settling over the earth. Slowly its shadow
blotted out the sun, and twilight engulfed the land while it
was still early afternoon. Mary Lizzie felt sure it was
the end of the world - that the sun would never shine again,
but it did. Slowly as the eclipse had shadowed the sun, it
passed over, and sunshine lifted the terrible gloom. Life
began again.

The first word of this terrible tragedy reached Lyman
while he was at work in the field. A passing neighbor
shouted to him, "Lincoln is assassinated!" He stopped
where he stood, unhitched his horses and drove them to
the barn as fast as he could, where he hitched onto the
buggy and drove into town, shouting the news to Jennett
as he left. He must learn more of the details of this
tragic event, all the time hoping that the rumor was un-
founded.

There was no more work done on the farm that day,
and through the slow hours of that sorrowful day there was
a general cessation of work. Factories, shops, stores
and mills were closed, while the entire country mourned
the passing of a great man. How could one work when
the President, our Beloved Lincoln, lay dead - and dead
by the hand of an assassin? The death of a thousand
Booths could not wipe out the stain; it was the foulest
crime the world had ever chronicled.

So passed into history Abraham Lincoln - life's greatest man. The years have placed him shoulder to shoulder with George Washington. Mary Lizzie ranks him above our beloved Washington - the greatest man that life has ever produced, in our own times or any other. Extravagant, perhaps, but honest.

The four terrible years of war were over; that great army of the North was disbanded, and its soldiers discharged; the boys were back again and at work on the farms. Life was slowly returning to normal; the women and children who had patiently carried on under the double load were now relieved of the strain; schools and colleges once more beckoned to the youth of the land and opened their doors in welcome; the years of almost deserted class rooms and empty halls were of the past, and soon forgotten. There would be many vacant desks, but life must go on, and work be planned to fill the gaps. New faces would occupy those places made vacant by the cruel war.

CHAPTER ELEVEN

The Curtis Farm

MANY visitors came to the farm that first year, for Grandmother Curtis still made it her home, and these visitors were largely relatives from the East who came to the farm for entertainment, as naturally as one would walk into the door of a hotel.

Nor did they fail of welcome. Lyman and Jennett were here as in every other home whole-hearted hosts, making welcome all who entered. Two brothers of Grandfather Newman came from Massachusetts with their wives that summer. They were followed by an aunt, a Mrs. Hale, and her daughter from Pittsfield, Massachusetts. The daughter, Carrie, was Mary Lizzie's age, and they became friends and corresponded for a number of years.

Katie and Mary Lizzie went to school in Otterville the first summer after the move to Independence, easily walking the two miles night and morning. Georgie, still Mary Lizzie's special charge, galloped gaily beside them on his stick to make his entrance into the school room, for he was now five years old.

Uncle Lew, a brother of Jennett's, and Aunt Hattie, his wife, were at the farm that summer, too. Their small son, Vernon, was Georgie's age, so Mary Lizzie had a team of two boys to manage, a task that sometimes taxed her powers heavily, for in. going to school they crossed a small stream, Harter's Creek, which showed a lovely expanse of clean white sand at the side of the bridge. It required skillful maneuvering to get the boys by in the morning with clean hands and faces for the school room. It did not matter so much at night; they could play then. In fact, Mary Lizzie joined them with gusto, for she too loved to dig in the sand.

Mary Lizzie takes the two small boys to school

After the year at Otterville, the new school house on the crossroad just west of the Curtis farm was finished, and Mary Lizzie attended here one year. Bert Lilly, Gent Sterns, his sister Abby, and one or two older boys from the outside also entered.

Her most vivid recollections center round the swimming lessons the big boys gave the two little fellows, her small brother Georgie, and Bert's brother Eddie, about the same age. A little creek had its birth in springs a

103

few miles to the east and north, and made its way across the border, or between the Lilly and the Curtis farms, passing a short distance from the school house. Its great attraction was a fine deep swimming hole, where the lessons were given.

Regularly during that early fall term the two small boys would be tossed by the big boys into the pool, and told they must swim to get out. Just as regularly, Mary Lizzie stormed and protested. She was sure her brother would go to the bottom some time, but it was astonishing how soon he and Eddie learned to strike out for shore.

In winter the creek and the swimming hole, with the gleam of ice that formed over the top, made a fine skating rink, and lessons in skating were given freely by those same big boys.

Mary Lizzie was growing up, and her father and mother thought it best to put her in a different atmosphere, for the section bordering on the farm was rather rough. In town was an excellent school, Miss Homan's and Mrs. Woodruff's Select School for Girls, and here she was entered. During the last years on the prairie Caro had lived the school months with Aunt Elinor Hart in Independence, and gone to this school which was recognized as of outstanding merit. From here she was sent to Upper Iowa University at Fayette, and she was there when the move to the Independence farm was made. That fall she remained at home, and again became a part of Miss Homan's School. These years belonged in the period before the public schools had reached their present high grade.

Will Wylie, son of a Bremer County neighbor, who was like an older brother in the family, drove the bay team, Betsy and Billy, before the democrat, leaving the two girls at Miss Homan's School while he went to an equally good school for young men.

This was the winter program. In the spring Caro remained at home, and Katie entered. She took her place in one of the small chairs at the front of Miss Homan's school room. Neither team nor driver could now be spared from the farm work, so Mary Lizzie and Katie walked to and from the school, two miles and a half each way. It was healthy exercise, and they flourished under it.

Caro was still the beauty of the family. Mary Lizzie was a slim, well-grown girl with heavy braids of long black hair and an olive skin. Her one redeeming feature was a fine straight nose. Mother Jennett and Caro were the fun makers of the family, and Mary Lizzie was the audience. Father Lyman was too busy, and Katie and Georgie were too young to linger around the after-dinner table for a merry chat as did these three, Mother Jennett, Caro and Mary Lizzie.

When there was no school, the children all had their part in the work of each day. Caro was her father's helper. In harvest she drove the reaper for him, while Mary Lizzie was at her mother's right hand until the inside work was done; then she took her share of picking up potatoes, dropping corn - just three, never more than four kernels in a hill, and in harvest time carrying bundles into place for Father Lyman to shock. Just ahead of her one day, the reaper stood still while the horses rested, and Caro slid down from her seat and started to walk around to her father. As she neared him she stepped squarely on a prairie rattler, fortunately just back of the venemous head, and so near it that it could not strike her, although the body thrashed instantly around her legs.

She emitted just one shriek and before the writhing body could curl down she sprang up her father's arm to his shoulder, feet drawn up and face buried in his neck. Equally quick her father's foot had descended on the hissing head, and that was the end of it - the reaping went on.

Mary Lizzie stepped high and watched for snakes as she carried her bundles that day. In fact a snake did once stick his head out of a bundle as she stopped to pick it up, but he was tied in so firmly that he could not escape. Mary Lizzie did not like this outdoor work. She preferred her dish washing, but it was necessary, for Lyman had no boys to help.

Uncle Will owned what was known as the 'Ide Farm,' one mile north of Lyman's and joining it. He was the joker of the Curtis family, and had merry blue eyes that twinkled in a face set in red hair and beard. His smile matched his twinkle and was quick and kindly.—He came down that first Christmas Eve at Independence just as the chickens were going to roost. A big gobbler had flown up on the roof and was poised on the chimney top. Uncle Will rushed in, gathered Georgie out of bed, and shouting, "I'll show you old Santa, Georgie," dashed outside and pointed to where the turkey roosted on the chimney, with, "There he is; do you see him?". Of course Georgie saw him, and kept his eyes so tightly closed lest Santa depart that, excited as he was, he was soon asleep.

The Lilly farm joined Father Lyman's on the north, and the two families were always pleasant friends. It happened once, when the Lilly family were away for the evening, that Katie had been asked to spend the night with Lizzie Lilly, the daughter who was about her age. The occasion was an entertainment in town that also took Father Lyman, Mother Jennett and Caro away from home. Mary Lizzie had been left with Georgie.

Late in the evening she thought she heard Katie sobbing at the front door. She flew to open it, and sure enough there stood poor little Kate, too much exhausted even to enter. She was minus cloak and hood, and was clutching her clothes with both hands to keep them on her body. When Mary Lizzie had gotten her warmed and soothed enough to talk coherently, she told a fearsome tale of robbers at the Lilly home.

The two children had started for bed and Katie was un-

dressed, when they heard a little noise. They waited only long enough for her to pull on her clothes again before
starting out to investigate. Suddenly the robbers were
there, and when one demanded of Lizzie where the money
was kept, Katie made a quick get away, cutting across the
farms, following the fences as she came. It was a night
of pitch blackness, and she could only feel, not see her
way. She was sure they had killed Lizzie. She had slipped out of the door just as she was, and had run all that
mile and a quarter home, following the fence. How she
had ever known to leave the fence when she came to the
creek and cross it on the bridge, then go back to the fence
again, seemed almost a miracle to Mary Lizzie, but she
had done it.

The two houses were alone that evening, and there was
no telephone, no near neighbor. Of course Mary Lizzie
thought a band of robbers would descend on her next. She
could not run away. She could only keep the two children
close to her and listen tensely while she waited for the attack to come. No sound had ever been so sweet to her as
her father's familiar "Whoa," when he drove into the yard.
She was safe, and Georgie and Katie were alive.

The Lilly team was not far behind, and Lyman notified
them of the trouble at their home and then followed them
to be on hand if help was needed. They found the house
rummaged from end to end and a badly frightened little
girl who had not been hurt by the robbers, and when they
left had undressed and gone bravely to bed.

The burglars were afterwards caught. They proved to
be men who had worked for Mr. Lilly and who knew the
lay of the house and Mr. Lilly's habit of keeping sums of
money at home. This happened to be one of the times
when there was little money to be stolen. The hired man
who was to have remained at the house that evening had
failed to do so. Although nothing was proved against him,
Mr. Lilly suspected that he was to share in the proceeds

and had helped to plan the robbery.

Mary Lizzie, grown into a slim straight girl of thirteen, in this year before she went away to school was handmaiden to the little withered Grandmother who sat patiently through many hours of the day with the big family Bible open on her lap. Occasionally Mary Lizzie was called upon to read from it, but memory did most of the reading, for the faded old eyes could not otherwise have followed even this big print.

Mary Lizzie loved books. In the years on the prairie her father's library held only a few volumes, but they were of the best: Shakespeare, Milton, Dante, Mrs. Hemans and Mrs. Segourney - two English poets - and Eugene Sue's 'Wandering Jew.' There were also two heavy leather bound volumes that held each a number of the shorter novels of the time. These she occasionally dipped into. While still quite a little girl she began to climb up and lift down one or the other of these volumes, puzzling over both words and meaning. She never read one book consecutively, but picked here and there, one author or another, as the fancy moved her.

Milton's 'Paradise Lost' held her attention longest, and perhaps she brought that volume down most often. Dante's 'Inferno' fascinated her, for she had a vivid imagination, and his horrors were terribly real. She dipped sometimes into 'The Wandering Jew.' Here it seemed to be the old white horse to which she turned most often. She grew to love the lofty style, high-flown phrases and perfect English. As she grew older, memory tucked away many beautiful things. Longfellow, Whittier, Holmes and Nathaniel P. Willis were absorbed by this voracious reader. Many of the shorter poems she could repeat at will.

Father Lyman always held season tickets for his whole family for a Lyceum Course every winter. Each evening of the Course was a treat. Mary Lizzie came to know personally men like Emerson, Longfellow, Whittier,

Holmes, Henry Ward Beecher, Theodore Tilton and J. G.
Holland. She shook hands with the author of 'Uncle Tom's
Cabin' and with Susan B. Anthony. These were not all.
The Swiss Bell Ringers took their turn with other sweet
singers. There was much of education as well as enjoy-
ment in the Lyceum Courses.

Later Mary Lizzie took her turn at Upper Iowa University, for a year and a half. This was a Methodist institu-
tion located at the little town of Fayette, some thirty miles
north of Independence, and was at that time the nearest
place for any advanced work educationally. Here she
passed the examinations for her first certificate and became a full fledged teacher.

Her first term of school was for six months, in the little school house just around the corner, an easy walk
from home. Following the close of that first term of
teaching, she went back to U. I. U. for the spring term.

The following winter she took her first heavy school in
the next district on the east. Here there was a large en-
rollment, and a long class - as they lined up - of young
men and young women, some of whom were older than the
teacher. They were forced to stand because of lack of
seating capacity, and she realized more fully how many
and how grown they were.

She worked hard, sometimes studying into the small
hours of the night to make sure that she knew each lesson
of the coming day, but she came through that winter with
flying colors. Six months at forty-five per made a big
check for Mary Lizzie to add to her school fund, and she
was mightily proud of it.

For the rest, there is little worth recording - just the
extra bits of life a girl has a right to: croquet, now and
then, an occasional spelling school or neighborhood party, sometimes a gay little buggy ride, sometimes a merry bunch of youngsters riding on a wide straw rack, and
in winter the big bob-sled with bells jingling merrily; oc-

casionally, very occasionally, for Father Lyman was rather particular, a ride in a fancy cutter behind a high stepping horse. Oh, then the bells jingled most merrily!

A ride in the Bobsled

CHAPTER TWELVE

College Years

College Years

Ring, College bell, old College bell
I fain would hear you call —
As when you rang for us
Each loiterer into class or hall

Ring, College bell, you grow not old
'Tis yours the charm of youth to hold
Your tunes are strong and clear and bold,
Ring on old bell, till time is told.
                                    L.C.F.

THE Iowa State Agricultural College was established in the fall of 1868, graduating its first class in 1872. In the beginning there was just one large building fronting east, made to house everything - office, chapel, classrooms and dormitories. Before that first class had been graduated, so urgent was the need for more room that long wings were extended backward; the old Main took the form of a hollow square open to the west.

The entrance to the campus was mid-way of the south side and the drive followed a graceful curve down and across a little canyon-like depression, where it divided; one part swept up past the site of the President's house and around in front of the main building, passing its full length, then on to curve around to the farm house; here it joined the other section and both ended at the barns.

The wings to the main building were already going up when Mary Lizzie entered College, and when they were finished the boys were housed on the third floor, while the second floor was given over to the public rooms, the teachers in charge, and the girls; and in spite of all its drawbacks it was a pleasant home.

The College year began in the spring, with graduation in the fall, so planned in order that vacation would come in the winter, when teachers commanded the highest salary. Here in March of '72, Mary Lizzie was to enter. She taught a four months' term of school that winter, but by March all was ready; a modest outfit, but enough. Another Independence girl was entering Ames that year, and the two were to be room-mates. Father Lyman went with them on that first trip. The College was so new, and so little was known of it, that I think he went to satisfy himself as to whether the change in institutions from U. I. U. to Iowa State had been a wise one. In those first years it was called the Iowa Agricultural College, and many were the jokes aimed at the 'Hay Seed College.'

Ames was well started, having an exceptionally fine faculty with President A. S. Welch at the head. Neither Mary Lizzie nor her father ever regretted the change; and she was always glad that she had her College years under President Welch.

The curriculum had been worked out by master minds; all study of the dead languages was excluded from the regular courses. A few terms of Latin were included as electives for the benefit of the English courses. It was, and is, a place to study the living languages, primarily the English. German and French were included, but the main emphasis was on literature and the business sciences; and Ames still ranks an exceptional school in these subjects.

The farm of 640 acres lay along one side of the railroad by which ran a public highway. The authorities had closed this road and opened a new one that skirted the other side of the College land; this meant that the road between Boone and Ames, of necessity, turned two almost right angled corners to get around the College land, of which this particular section had been reserved for the Campus, which had been carefully mapped out in accordance with that axiom of President Welch's that "curved is the line of beauty," and it held within its circle a wide sweep of lawn where occasional trees or groups of shrubs added to the beauty of the smooth greensward. All the public buildings were on the outside of that drive. A straight public road cutting that beautiful Campus in two could not be tolerated.

Naturally the public resented this round-a-bout road, and so besieged the County Commissioners that they ordered the College to take down its fences and let the public through. Instead of obeying this mandate, the College retaliated by having an especially strong fence built that would be hard to take down. Next, the Commissioners made public announcement that the road would be opened on a definite day, which happened to be a Saturday. If this was done, a public highway would cut through the center of the Campus - a thing that the College people would not for one moment tolerate.

President Welch called the boys together and discussed the situation with them; they agreed that the road must not go through. The boys were to remain at College all that eventful day. They had determined not to fight with the men sent by the authorities to take down the fence, but neither must the fence come down.

Acting under President Welch's advice, a few of the leading spirits organized the boys into a defensive army that was so completely to cover that section of the fence that the openers could not get hold of a single rail. A watch was posted in the College tower, and the bugle was to be blown at any suspicious appearance; the day passed without incident until late in the afternoon. Two or three of the boys who had been on guard duty thought it to good an opportunity for fun to be lost. They slipped away and approached the fence from the outside. To make it more realistic, as they came through Mr. Porter's farm yard, they picked up and carried over their shoulders axe, spade and shovel. The tower watchman saw them coming and thought the enemy was at hand; straightway the bugle blew long and loud, and the boys swarmed to protect that fence, covering it completely as well as the approach to it. Among them, his hat off and silvery locks blowing as wildly as any boy's, was President Welch, cautioning the crowd not to get excited, not to fight, but to hold that fence!

When the boys who had staged the attack saw how surprisingly successful had been their scheme, they disappeared and the defenders concluded that the enemy had flown. At any rate, the fence was saved, for the sun was setting and the day was at an end. Needless to say, it ended in an evening of general hilarity at the College. No other attempt was made to open the road, and the Campus was saved to become, in after years, one of the most beautiful in the United States. The joke was not known until long afterwards, and then only to a few; indeed it is doubtful if President Welch ever knew it was just College sport.

In March of 1872, Mary Lizzie, better known now as Lou Curtis, entered Iowa State College. She brought with her two special treasures: a Seth Thomas clock (a mantle one, in a mahogany case), and a hooked rug for her floor. The clock had timed her coming into the world in Kentucky, and had pioneered with the family into Iowa when she was a baby. For some reason, this clock was dubbed by her girl friends, "Lou Curtis's baby;" perhaps because she carried it in her arms whenever she went visiting. It counted her hours through four years of College, and graduated with her in 1875. Nor did it stop there; it timed her marriage, timed all her children,

114

timed the various family moves from Iowa, South Dakota, Montana, Utah and Wyoming into New Mexico. This last move proved too much for the faithful old time keeper. It stopped short, and now rests on a shelf in the New Mexico packing room. But sometime it is to be repaired so it may tick out its century of time.

Lou's baby

As for the rug, four years it lay on the floor of that College room, in front of the study table. On it fell many foot falls, and around it gathered many a council. The center of the rug held a broken basket, out of which tumbled three kittens. Mother Jennett had failed to hook in the fourth leg of the foremost little cat; what might have seemed a blemish was a special charm to the merry girls. No question could be settled until it had been argued out around that basket of kittens. It became the girls' call, "Let's go and talk it over with Lou's kittens," so into her room they would troop to sit on the floor in a circle around those little cats where the foremost one had but three legs.

A PIONEER GIRL

After the first year at College, Mary Lizzie would not
let Mother Jennett give it a fourth leg. Too many pow-
wows had resulted from that leg; so the three legged kit-
ten remained a sort of sign for the room. I regret to say
that when College days were over, Mother Jennett con-
signed that memorable rug, what was left of it, to the
refuse box.

Mary Lizzie maintains that the little three legged cat
was a powerful promotor in the change of her name from
Mary Lizzie to Lou. No College mate knew her except
as Lou. To her husband and her sisters she was, and is,
Lou, but to her father she was always Mary L., and her
mother, Lizzie, through all the years. Hers was no dual
personality; just a change in name that originated with
the Matron who had taken an especial liking to our Pioneer
Girl, and whom she thought resembled a former student
whose name was Lou; and the shorter name seemed to
fit the College girl.

In those days Iowa State required her students to work
a given length of time each day; the girls in the dormitories, and in the garden under the direction of the Professor of Horticulture; the boys doing all the farm work,
and such other duties as were considered too heavy for
the girls. The idea was that all the work about the College was to be done by the students.

Our girl Lou did her Freshman work in the kitchen -
washing dishes, setting tables or serving at the bread
counter. Her Sophomore year was in the chemical lab-
oratory, weighing out experiment materials and serving
the classes. Junior year found her assistant to one of the
Professors, whose office was in the Farm House, and as
a Senior she was given congenial work in the Library.
Here she was able to cultivate her taste for reading, and
broaden her knowledge of all class-room work.

The old College dining room was the scene of much
hilarity, and always of rush. From the time the 'jingle'

announced 'mess call' three minutes were allowed the
students in which to take their places, standing. Then
the doors were closed, silence was called, and General
Geddes, the College Steward, with raised hand and bowed
head, pronounced the blessing: "Bless, Oh Lord, these
mercies to our use; pardon our sins, for Christ's sake.
Sit down." Always the same, and always with a pause
before "Sit down!" Indeed, those irreverent boys and
girls insisted that the pause came before, and the clause
was, "For Christ's sake sit down." At any rate, that order, no matter where the pause came, was joyfully and
promptly obeyed.

One boy in the class possessed the ability to start a
laugh when and where he would. In the big College dining
room he would lay down his napkin, fold his hands, and
say with all seriousness, "Boys, it's time to laugh;" then
he would open his mouth and forth would peal a laugh so
hearty and infectious that one could not help joining. In
the fraction of a minute that whole dining room would be
laughing as though at some huge joke. No matter how
vigorously General Geddes might pound the floor order-
ing silence, the laugh kept going until Will stopped. It
left most of us breathless, and disgusted over our fool-
ishness, but laughing just the same.

Another prank was staged by a classmate - later to
become one of Omaha's millionaires, and the man who
was to figure most prominently in Lou's life and whose
name she later bore. Anniversary Day was a holiday,
and the students had hired all the livery rigs obtainable
to go in couples, or in groups to Boone or elsewhere,
searching a good time. The two boys, aided and abetted
by two fun loving girls, cleaned the farm dump cart, put
seats across it, harnessed a mule to it and, with many
flourishes, joined the procession of fine equipages. They
drove up in front of the main entrance, whose steps were
crowded with expectant students, dressed in their best

bib and tucker, and picked up those two jolly girls, who were dressed in gingham and sunbonnets, to match the boys' over-alls, the mule, and the dump cart. Many and loud were the cheers that greeted the turnout, and everywhere along the College roads they received the same acclaim. It was a bit of fun enjoyed by everyone.

When the Iowa State Oratorical Association was formed, Miss Curtis went as delegate from Ames to Grinnell, and had the honor of helping to draft the Constitution. Sororities and Fraternities were only just getting a start in the West; one Sorority had already established a Chapter at Mt. Pleasant, and likewise a Fraternity was formed there at the same time. Mt. Tabor, in southwestern Iowa, had a Chapter of another secret society for girls, the I. C.'s, so well established that it had begun reaching out to form new Chapters.

The week of graduation in the fall of '75, a representative of this latter Sorority was at State College to introduce them and a Chapter was then formed, which, however, Miss Curtis did not join as her time and attention were devoted to preparation for her future work. She already was engaged as assistant to the Principal of the Nevada High School; and her duties were to begin on the Monday following graduation.

Miss Curtis was an active member of the Philomathian Society, officiating during the four years from door keeper up to President. She also took part in College dramatics and was so realistic as a mad woman in 'Handy Andy,' tearing her long black hair and otherwise depicting that character, that she came near to being the 'Star.'

Her Junior oration brought her a compliment that I think pleased her better than any other that came to her in those days. The title of the oration was 'The Book of Revelations,' and when submitted to Professor Wynne, head of Literature, for criticism, he handed it back unchanged, with the comment, "It is beautiful, Miss Curtis, beautiful, beautiful."

Memory holds not only those words of commendation, through all the years, but a picture of Professor Wynne, as he turned his chair around, handing back the oration with one hand and stroking his long dark beard with the other.

Many more memories of those wonderful years are still vivid; for instance, the basket pie line (a cord outside the window, to lower a basket in which some kindly boy sent down a pie from Mrs. Porter's), or the merry dancers who escorted you down the hall, dancing a jig before you while they patted time with their hands, first in front, then behind, chanting,

> Rig-a-jig-jig, Rig-a-jig-jig,
> Gone to France to learn to dance,
> Rig-a-jig-jig!

One memory is of a pitiful night when she took her turn in keeping awake a poor girl who had, by mistake, taken an overdose of laudanum. Up and down the hall they paced, hour upon hour, now to stop at the open window, then again at the water faucet to dash cold water on the face of the sleeper, literally holding her up and forcing her to walk until the effects of the drug had worn off and her life was saved.

Happy memories of one merry Fourth of July, finished off with a midnight walk of two miles from Ames, where the buggies were left, back to College; a quiet entrance through the Secretary's office by means of a key carried in the pocket of one of the boys, a tip-toe up the back stairway, stealthily stealing past the Matron's door, down the long hall, and the safety of her own room at last. Oh, the fun of it, and Oh, the scandalized Matron!

One time an epidemic passed through the student body, and our girl contracted it and was forced to vacation home for two weeks. Gossip had it that the wave of sickness was instigated by College Authority on the supposition that it would brace the students up for hot weather. If it really was an experiment, the results were such that it was never tried again.

One year Mother Jennett came, unheralded, to visit her College girl. She arrived in Ames on the night train, which had no bus service. While she deliberated, uncertain what was best to do, two College boys who knew Lou Curtis offered to escort her to the College if she was able to walk so far. Unhesitatingly she accepted their offer, and proved to be such a good companion that she made friends of those two boys. They piloted her up the stairway, and down the long hall to her daughter's room. You can well imagine the joyful surprise it was to Lou, who had not dreamed of a visit from mother. Perhaps the Matron was somewhat scandalised to find that Lou's mother had come in so unceremoniously in the night, but what did that matter to the happy daughter?

Lou's Junior year found sister Kate, she of the mischievous dimple, enrolled as a student; later followed by George, the beloved little brother whom she had tended in childhood days. Both in turn were graduated from Iowa State.

College years were busy ones, yet there was always time for whatever came along. She represented Iowa State as a delegate to a conclave held at Des Moines, where arrangements for the first Interstate Debate were made. This was a pleasant change from College Dorm to the fine hostelry, 'Hotel Savory,' and was the means of making some very congenial acquaintances.

The College year closed in November with the usual Iowa dreariness of weather, but for the Class of 175, leaving I. A. C., there was nothing dreary about it; exams were all over, work practically all done. There remained but the finishing touches. Class Day exercises were held in the parlor of the Aborn House, Des Moines. Only twenty of that big Freshman class were on the final roster; some had fallen by the wayside, some had finished elsewhere, and four had crossed the river into Eternity; but lack of money had been the biggest factor in diminishing the number.

To Miss Curtis, the graduate, was given the honor of writing the Class Poem. This was a congenial task, so the work did not fall so heavily upon her in those last weeks.

Following are the opening stanzas of the poem the girl author produced:

> Old Time is a wondrous musician
> His chords are the swift changing years
> Each one in the anthem of ages
> With its measure of gladness or fears.
>
> The wierd relentless musician
> Chimes their changes, nor faster, nor slow
> Bear the chords joyous rhythm and gladness
> Or chimes in their grief, soft and low.

The week was full to the brim. On Sunday, President Welch's Baccalaureate Sermon was for the Class, to the, Class, and for once they gave close attention and listened with deepest interest. The Societies - Philomathian, oldest of all, gave them best wishes, and there were chats with the different Professors, and a last meeting of the Class.

Friends and relatives gathered to do honor to the graduates. Father Lyman, Mother Jennett and sister Caro were all there. Luther was there, and many of the lost ones who had started with 75 came back to clasp hands again. The week flew by, closing with the reception of honor, always given the graduating class by President and

Mrs. Welch.

After that, nothing remained but the hasty doing of forgotten things, the gathering up of luggage, the bus to Ames, the whistle of the incoming train. There was no time to be sad, just a lingering hand clasp and a parting God Speed. Life was all ahead.

CHAPTER THIRTEEN

Finis

Finis

The years of youth flow fast
So quickly girlhood sweeps along,
That e'er we know it, it is past.
So write it "Finis" for 'tis done.
              L.C.F.

WITH the passing of the thirteenth of November, 1875, College life was over. She was no longer Lou of College days, but Miss Curtis, the High School teacher. For the first time she was leaving College and beginning the year's work without a glimpse of home and a few days rest there. She was to be a teacher in the High School at Nevada, a small town near Ames.

A most convenient happenstance was that the new High School building was not to be ready for occupancy until the Monday after graduation, when the teacher also would be ready. It was a quick change from the gay excitement of the closing days at College to the staid decorum of the High School teacher; but Miss Curtis was used to quick changes, having throughout those four years taught a term of school beginning on the Monday after completing the College year, closing the four month term on Friday, and starting back to College early in the following week. She was always from one to two weeks late, but easily made up the back work.

As for getting ready, Mother Jennett and sister Caro took charge of that. The cleaning, mending, making and re-making of the College girl's clothing was carried on throughout the whole four months of her teaching, so that when her school term ended there remained but little to be done other than to close the register, send in her report and collect her salary. Her trunk was packed between whiles. This was the regular winter program, carried out with no thought that there could possibly be a failure anywhere along the line, nor was there.

Not even when she faced that large class of grown boys and girls, many of whom were older than herself, did any thought of fear cross her mind. She must admit however, that once when she had to take a half grown boy by the collar and jerk him up to attention, she found herself shaking after the ordeal was over - the boy having quickly complied with her order, although she did not recognize the spasm as fear. She has since wondered just what she would have done had that boy refused to obey. In all her teaching she never had occasion to use either ruler or switch. The year in Nevada passed quickly, her associations here being of the pleasantest, with Principal, teachers, pupils, boarding house and social life.

She went home for the Christmas vacation. Luther joined her there and the days flew by on wings; then no more vacationing, and, when the school year was finished, no more of teaching. Luther had waited patiently until her year was over, but that was to be the end of his waiting. July twentieth, the wedding anniversary of Father Lyman and Mother Jennett, was to be that of Luther and Lou, and there would be no postponement. Their first trip together was planned and carried out to the letter; this was to the wonderful Centennial Fair at Philadelphia, celebrating the One Hundredth Anniversary of the signing of the Declaration of Independence.

The twentieth fell on Thursday, and 'Thursday's the best day of all.' The wedding took place in the parlor of the house of seven gables, built for Grandfather Curtis

124

so long ago, two miles out from Independence on the Fair-
banks road. The ceremony was at noon, with only the
family present: sister Caro and her husband, George
Morse; Aunt Elinor Hart and the small Lillian; Uncle
Henry's wife, Addie, with her little daughter Florence,
and Earl Marsh, one of the College boys who was not only
a friend of Luther's, but of the family as well. Mother
Jennett's minister, Rev. Shafer of the Methodist Church,
married them. Lou wore her graduation gown, a silk
combined with silk poplin, in that soft shade of tan called
'Ashes of Roses.'

She came down the stairs with Sister Kate, who was
dressed in white swiss, and whom the minister took for
the bride-to-be.

Luther and the minister met them at the foot of the
stairs, and they passed into the parlor where the family
had congregated. The merry chat was stilled. Earl, as
'best man,' took his place at Luther's side, and sister
Kate, as bridesmaid, stood at the side of the bride, while
each answered in turn the age old question, "Will you take
this woman...this man...'til death do you part?   With
her hand in Luther's, she answered "Yes," even as he
had answered.

There was a short prayer, followed by the kisses and
best wishes of the family. One of Mother Jennett's famous
dinners followed, the table stretched to hold them all.

Afterwards she changed to the traveling gown of plain
gray poplin, donned the gray felt hat, its plainness relieved by a spray of purple velvet pansies, gathered up
her handbag and gloves, and was ready for the Great Adventure.

More kisses, but no tears, for this was a happy event.
Luther helped her into the democrat, and they were off,
Father Lyman taking them to the depot to catch the afternoon train for Philadelphia.   The future life stretched
ahead with the miles.

Tribute

...and so they were married, "and lived happily ever
after," or at least for 57 years together, making a genuine home full of love and understanding for their four
daughters, Florence Jennett, Clara Louise, Elizabeth
Curtis and Thora Lute.

In the many communities in which they lived, they
were always leaders in civic and educational advancement,
and their home was a social center for their friends - in
fact for the community.

Luther, always engaged in educational work, was a
great pioneer educator in the agricultural field, and for
many years was President of the New Mexico Agricultural
College.

Lou took her place in the cultural and social life of the
community. She was President of the Wednesday Liter-
ary Society, and first President of the American Association of University Women. She was honored by having
two of her poems, 'The Kneeling Nun' and 'The Organ
Mountains,' published in an Anthology of Contemporary
American Poets - 1929-1930.

They lived full lives together, respected and loved by
all who came in contact with them.

We, their daughters, have published PIONEER GIRL
so that the grandchildren may know something of their
fine heritage, and may have some of the pen pictures of
their beloved "Aunt Beth."

CLARA FOSTER BACON
LUTE FOSTER KISSAM

The War Songs

The War Songs, with their martial swing and beat, hold a place in history all their own. How rhythmically they flew along; how bravely they began:

> Rally round the flag, boys,
> Rally once again
> Shouting the battle cry of Freedom.
> The Union, forever,
> Hurrah, boys, Hurrah,
> Down with the traitor,
> And up with the stars.

> ### Chorus
> For we'll rally round the flag, boys,
> We'll rally once again,
> Shouting the battle cry of Freedom.

and then

> We do not want your cotton
> We care not for your slaves
> But rather than divide this land
> We'll fill your Southern graves.

> ### Chorus
> Hurrah! For equal rights, Hurrah!
> Hurrah for the brave old flag
> That bears the stripes and stars.

THE WAR SONGS

What a jolly swing there was to

When Johnnie comes marching home again,
Hurrah! Hurrah!
When Johnnie comes marching home again,
Hurrah! Hurrah!

and

We'll all drink stone blind
Johnnie fill up the bowl

When the call came for three hundred thousand more,
how quick was the response, and the songs followed just
as quick:

We are coming, Father Abraham
Three hundred thousand more
From Mississippi's winding stream
And from New England's shore.

Here the women singing:

Brave boys are they
Gone at their Country's call
And yet, and yet, we cannot forget
That many brave boys must fall.

Nor can we ever forget the pitiful cadence of 'The Vacant Chair.'

> We shall meet but we shall miss him
> There will be one vacant chair.
> We shall linger to caress him
> When we breathe our evening prayer.

Courage and help vibrated through every line of

> Tramp, tramp, tramp the boys are marching
> Cheer up comrades, they will come
> And beneath our starry flag
> We shall breathe the air again
> Of the freelands and our beloved home.

Who that lived through those years can ever forget the thrill of

> Tenting tonight on the Old Camp Ground
> Give us a song to cheer.

Or the prayer that breathes through

> Just before the battle, Mother
> I am thinking most of you
> While upon the field we're lying
> With the enemy in view.

Nor the piteous helplessness of the cry that follows:

THE WAR SONGS

Still upon the field of battle
I am lying, Mother, dear
With my wounded comrades waiting
For the morning to appear

Mother dear, your boy is wounded
And his night is full of pain
But still I pray that I may see you
And the dear old home again.

They sang snatches like this to keep up courage:

Oh, I'll eat when I'm hungry
I'll drink when I'm dry.
If the rebels don't get me,
I'll live 'till I die.

It is well to end these old War Songs with the triumph-
ant ring of 'Marching Through Georgia.'

Sherman's dashing Yankee boys
Will never reach the coast
So the saucy rebels said
And 'twas a handsome boast
Had they not forgot, alas
To reckon with the host
While we were marching through Georgia.

Chorus
Hurrah, hurrah, we bring the jubilee
Hurrah, hurrah, the flag that made you free
So we sang the Chorus from Atlanta to the sea
While we were marching through Georgia.